Abiotic Stress Effects on Performance of Horticultural Crops

Abiotic Stress Effects on Performance of Horticultural Crops

Special Issue Editors

Alessandra Francini
Luca Sebastiani

MDPI • Basel • Beijing • Wuhan • Barcelona • Belgrade

MDPI

Special Issue Editors

Alessandra Francini
Scuola Superiore Sant'Anna
Italy

Luca Sebastiani
Scuola Superiore Sant'Anna
Italy

Editorial Office
MDPI
St. Alban-Anlage 66
4052 Basel, Switzerland

This is a reprint of articles from the Special Issue published online in the open access journal *Horticulturae* (ISSN 2311-7524) from 2017 to 2019 (available at: https://www.mdpi.com/journal/horticulturae/special_issues/abiotic_stress).

For citation purposes, cite each article independently as indicated on the article page online and as indicated below:

LastName, A.A.; LastName, B.B.; LastName, C.C. Article Title. *Journal Name* **Year**, *Article Number*, Page Range.

ISBN 978-3-03921-750-2 (Pbk)
ISBN 978-3-03921-751-9 (PDF)

Cover image courtesy of Cristina Ghelardi.

Contents

About the Special Issue Editors

Alessandra Francini Ph.D. studies the stress biology of woody and herbaceous crops. She has been working as a Laboratory Technician at BioLabs, Institute of Life Science, Scuola Superiore Sant'Anna, Pisa, Italy, since 2019. Her main research topics include antioxidant compounds in plants; chlorophyll *a* fluorescence analyses; fruit quality; heavy metal pollution; phytoremediation; and xenobiotic metabolism in woody and fruit trees.

Luca Sebastiani is Full Professor, Horticultural Sciences, Section of Agricultural Science and Biotechnology, Institute of Life Sciences, Scuola Superiore Sant'Anna. He has been serving as Director of the Institute of Life Sciences, Scuola Superiore Sant'Anna, since his appointment in June 2016. His main research topics of interest include physiological and molecular interactions among plants and environmental constrains; fruit composition and antioxidant compounds; and woody plant germplasm conservation ICT and DSS technologies in agriculture.

horticulturae

MDPI

Editorial

Abiotic Stress Effects on Performance of Horticultural Crops

Alessandra Francini * and Luca Sebastiani

BioLabs, Institute of Life Science, Scuola Superiore Sant'Anna, Piazza Martiri della Libertà 33, I-56127 Pisa, Italy; l.sebastiani@santannapisa.it
* Correspondence: a.francini@santannapisa.it

Received: 17 September 2019; Accepted: 19 September 2019; Published: 26 September 2019

Abstract: The yield and quality of horticultural crops mainly depend on genotype, environmental conditions, and cultivation management. Abiotic stresses, such as adverse environmental conditions, can strongly reduce crop performance, with crop yield losses ranging from 50% to 70%. The most common abiotic stresses are represented by cold, heat, drought, flooding, salinity, nutrient deficiency, and high and low light intensities, including ultraviolet radiation. These abiotic stresses affect multiple physiological and biochemical processes in plants. The ability of plants to face these stresses depends on their adaptation aptitude, and tolerant plants may express different strategies to adapt to or avoid the negative effects of abiotic stresses. At the physiological level, photosynthetic activity and light-use efficiency of plants may be modulated to enhance tolerance against the stress. At the biochemical level, several antioxidant systems can be activated, and many enzymes may produce stress-related metabolites to help avoid cellular damage, including such compounds as proline, glycine betaine, amino acids, etc. This special issue gathers eight papers; three are reviews and five are research papers. Two reviews are focused on the application of appropriate agronomic strategies for counteracting the negative effects of abiotic stresses. The third review is based on ornamental plant production under drought stress conditions and the effect on their ornamental quality. The research papers report the effect of climate change on crop development, yield, and quality. Abiotic stresses have been proven to reduce crop performance and yield. Research studies are essential for understanding the key adaptation strategies of plants that can be exploited for improving the crop stress tolerance.

Keywords: agronomic tools; dormancy; drought; ornamental; salinity

---

1. Introduction

Environmental stresses are the main factor limiting production in agricultural systems. Abiotic stresses, such as adverse environmental conditions, can strongly increase crop yield losses (Figure 1), ranging from 50% to 70% [1]. Climate change is often mentioned as one of the future challenges that the agricultural sector must face. An increase in temperature is considered the crucial factor that will reduce water quantity and quality. The availability of this natural resource affects the lives of human beings, as well as agricultural yields. Fresh water is an essential resource for ecosystems and humanity. The use of water is under increasing pressure in many parts of the word, particularly for agriculture, which is by far the largest water-use sector, accounting for around 70% of the water withdrawal worldwide [2]. Climate change requires a more sustainable management of water resources. However, in several agricultural areas, excessive rain can also induce flooding, with negative effects on crop development and production. In these areas it is important to select suitable crops and control soil water though appropriate agronomic management strategies.

Besides high temperatures, low temperatures can also represent a problem for many macrothermal species that are grown in spring, or fruit trees that have their bloom period in the early spring (such as peaches or almonds). Cold stress can induce chilling injury and damage flowers or leaves with subsequent significant yield losses.

High or low irradiance can also have direct impacts on crop biomass and produce quality since photosynthesis and respiration are tightly correlated with temperature, light intensity, and quality.

Inadequate crop nutrition can induce excesses or lack of some essential elements, with effects on growth, yield, or quality of the produce depending on the species. The excess of heavy metals or xenobiotic compounds can also represent a limiting factor in different urban and agricultural areas. This special issue collects state of the art studies of the effects of abiotic stresses on agricultural crops and ornamental plants in Mediterranean areas. In particular, a compendium of agronomic strategies has been described and re-evaluated for improving crop tolerance in the short term [3,4].

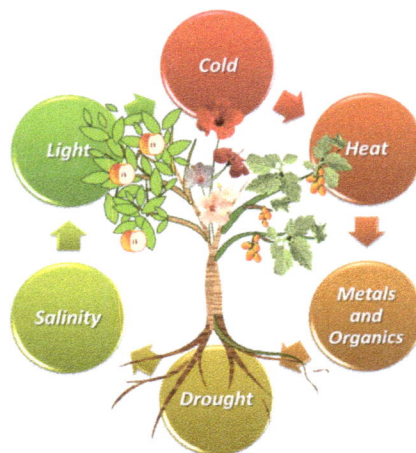

Figure 1. Potential abiotic stresses that can compromise crop yield and produce quality.

2. Controlled Abiotic Stress Management for the Agriculture Production

The agricultural market is constantly oriented to produce the most common crops year-round, or to exploit the lower market availability of some products in early spring or late winter for getting the highest prices. The out-of-season production is often performed in greenhouses and requires high energy consumption. Therefore, suboptimal temperatures or light conditions can represent important factors to manage for avoiding crop damage and excessive production costs. Experimental work was performed in bedding plant production in greenhouses during winter, with exposure to low-energy conditions characterized by reduced temperature and light conditions for a two-week period over a growing cycle of eight weeks. Results showed negative effects on flowering and plant growth by the addition of a two-week low-energy exposure. In particular, flowering was delayed, and reductions in flower number, plant size, and biomass were observed. The most affected crops where those that were cold-sensitive, such as impatiens [5]. Further, studies should be carried out on cold tolerant species and in non-flowering species.

Environmental parameters have direct effects on crop performance in different seasons and different nutrient availability. Cultivation carried out with two different lettuce cultivars in different seasons and with different nutrient availabilities showed that suboptimal growing conditions limit nutrient utilization and have effects on biomass accumulation. Secondary metabolites, which can contribute to the antioxidant capacity of lettuce, were affected by the seasons by effects on both composition of different flavonoids and in their total concentrations [6]. Considering the overall results, higher nutrient solution concentrations should be used in spring for maximizing yield and quality in lettuce.

The yield of crops is directly correlated with photosynthesis and the main factors involved in this physiological process. Water is an essential element of photosynthesis, and its availability can be directly correlated with yield and quality. Prediction models have been developed for estimating yield in different stress conditions. In this special issue, a prediction model based on evapotranspiration has been used for estimating the yield of apple under water deficit conditions [7]. The model was studied for apple yield estimation under three cultivation regimens: conventional irrigation, partial root zone drying, and continuous deficit irrigation. Results showed that the model worked well for vigorous cultivars such as 'Fuji', while it did not perform well for cultivars like 'Gala' that are not able to limit water losses by closing stomates. In pepper, it was demonstrated that salinity and water availability affected the yield and quality at harvest and during postharvest storage [8].

The adaptation of crops to stressful conditions can be achieved through the selection of a suitable genotype with specific traits. Genetic improvement programs can be used for enhancing tolerance to the different stresses, but require long periods of work. In the short term, agronomic strategies can be adopted for reducing stress intensity to the crops. A compendium of old and new agronomic tools has been reported in this special issue. For each stress, specific agronomic strategies have been described for lowering their negative effects and allow crops to cope with the stressful conditions [3,4].

3. Cold Stress and Bud Dormancy Transition

Abiotic stresses may also be utilized in some species for synchronization with the seasonal change. In many fruit tree species, cold stress and the accumulation of cold units are essential for bud differentiation. In this special issue endodormancy of almond and a putative regulatory gene, the *Dormancy Associated MADS*-Box (DAM), has been studied [9]. Since it is well known that temperature trends affect bud dormancy, in this study the expression of *PdDAM6* was compared in warmer and colder seasons. Results indicated that the endodormancy to ecodormancy transition involved a transcriptional reprogramming, in which genes acting on dormancy maintenance would be downregulated. In almond, the expression of *PdDAM6* seemed to play a crucial role.

4. Drought Stress and Ornamental Plants

The quality of ornamental plants depends on their visual appearance, which is defined by such factors as leaf color, size, number, and longevity. Abiotic stresses and in particular drought stress can severely affect leaf morphology and physiology during adaptation to stressful environments. These responses can have a direct impact on ornamental quality and subsequently on the commercial value of the plants. A review included in this special issue describes the physiological, biochemical, and morphological changes that ornamental plants can undergo under drought stress, and how these influence quality [10]. The most common changes that can be observed on leaves are smaller size and their orientation on the branch. Ornamental plant drought stress responses are important for their selection in relationship to their area of utilization, such as urban or peri-urban areas.

5. Conclusions

Abiotic stresses have been proven to reduce crop performance and yield. However, mild stresses can also have positive effects on the quality of produce, especially through the activation of the phenylpropanoid pathway and the accumulation of bioactive compounds. These can improve postharvest performance and enhance the nutritional quality of the produce, which is particularly important for the consumer.

Abiotic stresses must be continuously studied with multidisciplinary approaches, from the basic science for understanding crop responses, and their adaptation to the identification of practical agronomic solutions for alleviating the stressful effects and preserving crop productivity.

Funding: This research received no external funding.

Acknowledgments: We gratefully acknowledge all the Authors that participate to this Special Issues.

Conflicts of Interest: The authors declare no conflict of interest.

References

1. Boyer, J.S. Plant productivity and environment. *Science* **1982**, *218*, 443–448. [CrossRef] [PubMed]
2. FAO. *Coping with Water Scarcity. An Action Framework for Agriculture and Food Security*; FAO: Rome, Italy, 2008; p. 100.
3. Mariani, L.; Ferrante, A. Agronomic Management for Enhancing Plant Tolerance to Abiotic Stresses—Drought, Salinity, Hypoxia, and Lodging. *Horticulturae* **2017**, *3*, 52. [CrossRef]
4. Ferrante, A.; Mariani, L. Agronomic Management for Enhancing Plant Tolerance to Abiotic Stresses: High and Low Values of Temperature, Light Intensity, and Relative Humidity. *Horticulturae* **2018**, *4*, 21. [CrossRef]
5. Boldt, J.K.; Altland, J.E. Timing of a Short-Term Reduction in Temperature and Irradiance Affects Growth and Flowering of Four Annual Bedding Plants. *Horticulturae* **2019**, *5*, 15. [CrossRef]
6. Sublett, W.L.; Barickman, T.C.; Sams, C.E. The Effect of Environment and Nutrients on Hydroponic Lettuce Yield, Quality, and Phytonutrients. *Horticulturae* **2018**, *4*, 48. [CrossRef]
7. Lo Bianco, R. Water-Related Variables for Predicting Yield of Apple under Deficit Irrigation. *Horticulturae* **2019**, *5*, 8. [CrossRef]
8. Fallik, E.; Alkalai-Tuvia, S.; Chalupowicz, D.; Zaaroor-Presman, M.; Offenbach, R.; Cohen, S.; Tripler, E. How Water Quality and Quantity Affect Pepper Yield and Postharvest Quality. *Horticulturae* **2019**, *5*, 4. [CrossRef]
9. Prudencio, Á.S.; Dicenta, F.; Martínez-Gómez, P. Monitoring Dormancy Transition in Almond [*Prunus dulcis* (Miller) Webb] during Cold and Warm Mediterranean Seasons through the Analysis of a DAM (Dormancy-Associated MADS-Box) Gene. *Horticulturae* **2018**, *4*, 41. [CrossRef]
10. Toscano, S.; Ferrante, A.; Romano, D. Response of Mediterranean Ornamental Plants to Drought Stress. *Horticulturae* **2019**, *5*, 6. [CrossRef]

horticulturae

MDPI

Article

Timing of a Short-Term Reduction in Temperature and Irradiance Affects Growth and Flowering of Four Annual Bedding Plants

Jennifer K. Boldt * and James E. Altland

United States Department of Agriculture, Agricultural Research Service, Application Technology Research Unit, Wooster, OH 44691, USA; James.Altland@ars.usda.gov
* Correspondence: Jennifer.Boldt@ars.usda.gov; Tel.: +1-419-530-2225

Received: 7 December 2018; Accepted: 23 January 2019; Published: 1 February 2019

Abstract: Heating and supplemental lighting are often provided during spring greenhouse production of bedding plants, but energy inputs are a major production cost. Different energy-savings strategies can be utilized, but effects on plant growth and flowering must be considered. We evaluated the impact and timing of a two-week low-energy (reduced temperature and irradiance) interval on flowering and growth of impatiens (*Impatiens walleriana* Hook.f. 'Accent Orange'), pansy (*Viola × wittrockiana* Gams. 'Delta Premium Blue Blotch'), petunia (*Petunia × hybrida* Hort. Vilm.-Andr. 'Dreams Pink'), and snapdragon (*Antirrhinum majus* L. 'Montego Violet'). Flowering was delayed 7 to 10 days when the low-energy exposure occurred before flowering. Flower number was reduced 40–61% in impatiens, 33–35% in petunia (low-energy weeks 5–6 and weeks 7–8, respectively), and 35% in pansy (weeks 5–6). Petunia and impatiens dry mass gradually decreased as the low-energy exposure occurred later in production; petunias were 26% (weeks 5–6) and 33% (weeks 7–8) smaller, and impatiens were 20% to 31% smaller than ambient plants. Estimated energy savings were 14% to 16% for the eight-week period, but only up to 7% from transplant to flowering. Growers can consider including a two-week reduction in temperature and irradiance to reduce energy, provided an additional week of production is scheduled.

Keywords: temperature; irradiance; ornamental plants; greenhouse production

---

1. Introduction

Greenhouse production of annual bedding plants for spring markets occurs in late winter and early spring. Heating and supplemental lighting are often provided to offset low outdoor temperatures and augment low solar irradiation intensity and duration. Consequently, energy inputs are a major production cost for greenhouse-grown plants. In the United States, energy is the third largest expense after labor and plant material, and it accounted for 9% of total production costs in the 2014 Census of Horticultural Specialties [1]. Approximately 65% to 85% of total greenhouse energy consumption is for heating [2].

Irradiance drives photosynthesis and primarily influences crop growth and dry weight gain. Temperature primarily influences crop development, including rates of leaf unfolding, flower initiation, and flower development. Together, light and temperature impact crop timing and quality. The ratio of radiant energy (light) to thermal energy (temperature), or RRT, is one way to describe this [3]. A higher RRT increases crop quality; for example, plants grown at lower temperatures and higher irradiance (high RRT) will be of higher quality than plants grown at higher temperatures and lower irradiance (low RRT). This ratio is an indicator of plant carbon balance, which becomes depleted under prolonged exposure to high temperature and low irradiance [4]. For example, starch levels in rose (*Rosa × hybrida*

'Red Berlin') were similar when plants were exposed to high temperature and high irradiance or low temperature and low irradiance but were diminished under high temperature and low irradiance [4].

Crops need to meet both a target market date and minimum quality standards. Lighting and temperature set points can be adjusted during production to address economic and environmental concerns. Growers are continually looking for strategies to reduce energy consumption and costs without sacrificing plant growth, quality, and/or finished time. Therefore, it is important to evaluate different production strategies and determine when they may be feasible to implement. Surveys have found 55% to 58% of responding U.S. growers have implemented conservation or energy efficient practices [5,6]. Installing an energy or thermal screen is one strategy for reducing energy use [2], but only 12% of respondents in one of the surveys had them [5]. Other viable production strategies for reducing energy use include growing in unheated greenhouses or high tunnels [7], using root zone heating [8], or using a reduced temperature to finish (RTF) [9].

Many growers maintain static air temperature set points, independent of ambient weather conditions outside [10]. Allowing the greenhouse temperature to rise above the desired mean daily temperature (MDT) when heating demand is low and fall below the desired MDT when heating demand is high, but maintain the same MDT, is another strategy to reduce energy use [11]. These temperature integration strategies have been referred to as dynamic temperature control [2], integrating temperature control [12], multi-day temperature setting [13], or dynamic photosynthetic optimization [14]. While MDT remains the same, a wider range of acceptable temperature fluctuations is allowed. This can be accomplished by increasing the daytime ventilation set point and decreasing the nighttime heating set point, using a computer algorithm to maintain a rolling MDT and adjust temperatures based on predicted weather patterns [15], or using a computer algorithm to adjust greenhouse conditions based on photosynthetic optimization [14,16] or plant assimilate balance [13]. Temperatures need to remain within the linear range of plant development rate, between the base temperature (T_{base}; development rate = 0) and the optimum temperature (T_{opt}; development rate is maximal) for each species to minimize delays in development [17].

Dynamic temperature management can be integrated on a 24 h [11,18], multi-day [15], or weekly basis [19,20]. It has been successful for roses [12,21], potted plants [18,22], and vegetables [19] when the amplitude of the bandwidth was $\leq \pm 6\ ^\circ C$. One drawback, however, is that dynamic heating requires a greenhouse environmental control computer with sophisticated software, and not all growers have or can afford these systems [2]. For greenhouses without environmental control systems, an energy-reduction alternative is to optimize the growing environment on days with a lower heating requirement (i.e. warmer, sunnier, and or less windy) and reduce energy inputs (lower the temperature, turn off supplemental lighting, and close the energy curtain) on days that require more heating. This short-term reduction in temperature and irradiance was successful when implemented 1 to 2 days per week, with reduced energy costs and minimal impact on plant growth and crop timing [23,24]. However, weather patterns often are cyclical on a longer time scale, and growers may contend with days to weeks of continuous cloudy weather during winter and early spring. Condensing the low temperature and irradiance exposure into a continuous time period rather than interspersed throughout production may impact its successful implementation, even though it is for a similar total number of days. Therefore, our objective was to evaluate the timing of a two-week reduction in temperature and irradiance on plant growth and flowering of four popular annual bedding plant crops and estimate potential cost savings using the Virtual Grower software program. We selected cold-tolerant [pansy (*Viola* × *wittrockiana* Gams.) and snapdragon (*Antirrhinum majus* L.)], cold-intermediate [petunia (*Petunia* × *hybrida* Hort. Vilm.-Andr.)], and cold-sensitive [impatiens (*Impatiens walleriana* Hook.f.)] species, as categorized by their T_{base} [25,26].

2. Materials and Methods

Seeds of impatiens 'Accent Orange', pansy 'Delta Premium Blue Blotch', petunia 'Dreams Pink', and snapdragon 'Montego Violet' were sown on 15 December 2014 (replication 1) and 6 January

2015 (replication 2) into 288-cell plug trays filled with a peat-based soilless substrate (LC-1; Sun Gro Horticulture, Bellevue, WA). Trays were placed in a growth chamber (GR48; Environmental Growth Chambers, Chagrin Falls, OH) set to provide 25 °C constant air temperature, 300 $\mu mol \cdot m^{-2} \cdot s^{-1}$ photosynthetic photon flux density (PPFD) from high-pressure sodium (HPS) lamps, and an 8 h photoperiod. They were watered as needed, and provided 75 $mg \cdot L^{-1}$ N constant liquid feed of 20N–4.4P–16.6K (Jack's 20-10-20; JR Peters, Inc., Allentown, PA, USA) at each irrigation once true leaves emerged.

Two greenhouse environments were set up in identical compartments located within a glass-glazed greenhouse (Toledo, OH, USA). The ambient compartment represented typical greenhouse conditions during winter and early spring. Temperature set points were 22 °C day/18 °C night. High-pressure sodium light fixtures (Sunlight Supply, Inc., Vancouver, WA, USA) provided approximately 75 $\mu mol \cdot m^{-2} \cdot s^{-1}$ of supplemental irradiance from 1000-W bulbs (Osram Sylvania Products, Inc., Manchester, NH, USA) when ambient PPFD at the benchtop was less than 300 $\mu mol \cdot m^{-2} \cdot s^{-1}$, and a constant 14 h photoperiod (0600-2000 HR) was maintained. The cool compartment represented a cool, low light environment. Temperature set points were 13 °C day/10 °C night. A spun-woven energy curtain was continuously closed, which reduced ambient irradiance by approximately 50%, relative to ambient conditions. Day-extension lighting with HPS lamps was provided when ambient irradiance was less than 10 $\mu mol \cdot m^{-2} \cdot s^{-1}$ PPFD to achieve a constant 14 h photoperiod. Dataloggers (HOBO®Pro v2; Onset Applications, Bourne, MA, USA) in each environment measured air temperature. Quantum sensors (Model QSO-S; Apogee Instruments, Logan, UT, USA) were connected to a data logger (CR10X; Campbell Scientific, Logan, UT, USA) and mean PPFD was recorded every 15 min.

On 20 January and 9 February 2015, plants were transplanted into 11.5 cm diameter round pots filled with LC-1. They were irrigated as needed during the experiment. Plants were watered once weekly with reverse-osmosis water and fertilized with 20N–4.4P–16.6K at an N concentration of 150 $mg \cdot L^{-1}$ at all other irrigation events. Electrical conductivity (EC) and pH of the substrate solution was monitored every two weeks on three additional plants of each species grown continuously in the ambient and cool, low light environments, using the Pour-Through technique, to ensure values remained within the recommended ranges for all species [27].

Plants were moved from ambient conditions to the cool, low light environment for a two-week interval during the eight-week duration of the experiment (i.e., weeks 1–2, weeks 3–4, weeks 5–6, or weeks 7–8 in cool, low light, with the other 6 weeks in ambient conditions). Two additional treatments included a continuous ambient control and a continuous cool, low light control. There were five plants per species per treatment. Mean air temperatures and daily light integrals (DLIs) are provided in Table 1.

Table 1. Mean air temperature (°C) and daily light integral (DLI, $mol \cdot m^{-2} \cdot d^{-1}$) for each treatment. Plants were grown in ambient conditions [22/18 °C air temperature, 14 h photoperiod, and ambient irradiance +75 $\mu mol \cdot m^{-2} \cdot s^{-1}$ supplemental lighting from high-pressure sodium (HPS) lamps when photosynthetic photon flux density (PPFD) was less than 300 $\mu mol \cdot m^{-2} \cdot s^{-1}$] and transferred to cool, low light conditions (13/10 °C air temperature, ambient irradiance, and a 14 h photoperiod achieved by providing 75 $\mu mol \cdot m^{-2} \cdot s^{-1}$ from HPS lamps when PPFD was less than 10 $\mu mol \cdot m^{-2} \cdot s^{-1}$) at 2-week intervals during an 8-week production cycle.

Treatment	Target Mean Temperature (°C)	Mean Temperature (°C)		DLI ($mol \cdot m^{-2} \cdot d^{-1}$)	
		Replication 1	Replication 2	Replication 1	Replication 2
Ambient	20.3	20.3 ± 1.7	20.3 ± 1.7	10.7 ± 3.0	12.8 ± 3.4
Weeks 1–2 cool	18.2	18.3	18.7	10.2	11.6
Weeks 3–4 cool	18.2	18.5	18.3	9.7	11.3
Weeks 5–6 cool	18.2	18.5	18.6	9.4	10.9
Weeks 7–8 cool	18.2	18.5	18.4	8.9	10.7
Cool	11.8	12.9 ± 1.1	13.2 ± 1.4	6.1 ± 1.5	6.3 ± 1.9

Flowering was checked daily, and date of first flower was recorded. Eight weeks after transplant, flower number was counted. Relative chlorophyll content (CCM-200; Apogee Instruments, Inc., Logan, UT) was measured on three recently mature leaves per plant, and the mean value was used for statistical analysis. Plant height was measured from the substrate surface to the apex. Plant width was measured at the widest point and perpendicular to the widest point, then the two measurements were averaged. Above-ground plant tissue was removed, washed with 0.1 N HCl, rinsed with ultra-purified (18 MΩ) water, dried in a forced-air oven at 60 °C for 3 days, and weighed for dry mass.

A virtual greenhouse was constructed to estimate daily heating and supplemental lighting costs for each environment, using the USDA-ARS software program Virtual Grower 3.0.9 (USDA-ARS, Toledo, OH, USA). Greenhouse dimensions, materials, and components were as described previously [24]. Total energy costs for each treatment were calculated (1) as the sum of daily energy costs for the eight-week production duration, and (2) as the sum of daily energy costs from the start of the experiment to mean date of the first flower, based on the temperature and supplemental lighting schedules.

Data were analyzed as a randomized complete block design, with six treatments, five single-plant replications per treatment, and repeated twice in time. A separate analysis was conducted for each species. Data were analyzed in SAS (SAS 9.3; SAS Institute, Inc., Cary, NC, USA) using the GLM procedure (PROC GLM) and mean separation was conducted with Tukey's HSD at $\alpha = 0.05$ for significant treatment effects ($P \leq 0.05$).

3. Results and Discussion

Length of production is a critical benchmark for greenhouse growers, as increased production time reduces the number of crop cycles per season and increases the fixed costs allocated to each crop. Flowering is important for quick sell-through at retail, and therefore, time to flower is an important scheduling metric. Compared to control plants grown at ambient conditions, flowering was delayed when plants were provided with a two-week low-energy interval early in production. This occurred in all crops grown in the cool, low light environment in weeks 1–2 or weeks 3–4, and additionally in petunia in weeks 5–6 (Table 2). The delay in flowering was 7 to 10 days. Ambient control plants began flowering during week 5 of production (29 to 35 days after transplant); mean days to flower was 30, 32, 34, and 34 days in pansy, impatiens, snapdragon, and petunia, respectively. As such, the absence of delayed flowering on plants exposed to the low-energy exposure in weeks 5–6 or weeks 7–8 occurred because plants were budded or flowering before the start of the cool, low light exposure.

Continuous exposure to low-energy conditions delayed flowering by more than 3 weeks, compared to ambient controls (Table 2). Incomplete flowering occurred in impatiens and snapdragon by the end of the experiment, 8 weeks after transplant, and none of the petunias had flowered. Minimum temperature (T_{min}) values for flower development of 2.0 to 4.0 °C have been reported for snapdragon, 2.8 to 5.5 °C for petunia, 4.1 °C for viola (*Viola cornuta* L. 'Sorbet Plum Velvet'), and 7.2 °C for impatiens 'Blitz 3000 Deep Orange' [25,26]. Although we used different cultivars in our study, the 10 °C nighttime set point in the low-energy treatment was greater than the T_{min} for all species. Therefore, flowering would have eventually occurred, but the delay and variability of flowering does not make continuous low-energy conditions a viable production strategy for most species.

Table 2. Plant growth and flowering of impatiens (*Impatiens walleriana* 'Accent Orange'), pansy (*Viola* × *wittrockiana* 'Delta Premium Blue Blotch'), petunia (*Petunia* × *hybrida* 'Dreams Pink'), and snapdragon (*Antirrhinum majus* 'Montego Violet') grown in ambient conditions (ambient), grown in ambient conditions and transferred to a low-energy environment at two-week intervals during an eight-week production cycle, or grown continuously in the low-energy environment (continuous). Ambient conditions were 22/18 °C air temperature, 14 h photoperiod, and ambient irradiance + 75 $\mu mol \cdot m^{-2} \cdot s^{-1}$ supplemental lighting from high-pressure sodium (HPS) lamps when photosynthetic photon flux density (PPFD) was less than 300 $\mu mol \cdot m^{-2} \cdot s^{-1}$. Low-energy conditions were 13/10 °C air temperature, ambient irradiance, and a 14 h photoperiod achieved by providing 75 $\mu mol \cdot m^{-2} \cdot s^{-1}$ from HPS lamps when PPFD was less than 10 $\mu mol \cdot m^{-2} \cdot s^{-1}$.

Crop	Treatment	CCI (4 Weeks after Transplant)	CCI (8 Weeks after Transplant)	Height (cm)	Width (cm)	Dry Mass (g)	Flower Number	Days to Flower
Impatiens	Ambient	55.5 ± 3.1	56.7 ± 3.4	19.9 ± 0.7	42.1 ± 1.4	14.9 ± 0.6	58.2 ± 5.9	32 ± 2
	Weeks 1–2	37.9 ± 1.5	57.6 ± 2.9	18.4 ± 0.9	38.4 ± 0.9	11.9 ± 0.5	34.8 ± 2.6	43 ± 2
	Weeks 3–4	30.2 ± 1.5	51.4 ± 2.7	16.8 ± 0.8	35.8 ± 1.4	10.9 ± 0.6	22.9 ± 2.2	42 ± 2
	Weeks 5–6	57.4 ± 3.1	59.5 ± 3.6	18.0 ± 1.0	36.2 ± 1.5	10.9 ± 0.3	28.3 ± 2.6	37 ± 3
	Weeks 7–8	57.1 ± 3.3	53.5 ± 4.0	16.0 ± 0.9	34.3 ± 1.4	10.3 ± 0.6	26.8 ± 4.6	38 ± 2
	Continuous	13.1 ± 1.0	34.7 ± 1.4	9.9 ± 1.0	15.1 ± 0.8	1.1 ± 0.1	0.3 ± 0.2	54 ± 2
	ANOVA [z]	<0.0001	<0.0001	<0.0001	<0.0001	<0.0001	<0.0001	<0.0001
	HSD$_{0.05}$ [y]	10.3	13.3	2.8	4.0	2.2	14.8	8
Pansy	Ambient	75.4 ± 1.3	60.8 ± 4.1	12.3 ± 0.6	16.6 ± 1.1	3.3 ± 0.4	11.1 ± 0.9	30 ± 1
	Weeks 1–2	61.5 ± 2.8	63.5 ± 2.3	13.0 ± 0.6	17.7 ± 1.2	3.6 ± 0.6	8.9 ± 0.7	38 ± 2
	Weeks 3–4	54.1 ± 2.5	65.3 ± 4.3	12.1 ± 0.7	17.5 ± 0.9	3.7 ± 0.3	8.8 ± 1.3	38 ± 2
	Weeks 5–6	71.2 ± 3.2	74.7 ± 3.3	11.5 ± 0.6	17.2 ± 0.9	3.9 ± 0.5	7.2 ± 1.0	37 ± 3
	Weeks 7–8	65.7 ± 4.2	60.6 ± 3.0	14.1 ± 0.4	18.8 ± 0.9	3.4 ± 0.3	8.0 ± 0.8	33 ± 2
	Continuous	43.1 ± 2.3	67.3 ± 3.1	11.2 ± 0.8	15.3 ± 0.4	1.8 ± 0.1	0.9 ± 0.2	54 ± 1
	ANOVA	<0.0001	0.0505	0.0051	0.0211	<0.0001	<0.0001	<0.0001
	HSD$_{0.05}$	12.1	-	2.3	2.9	1.1	3.6	7
Petunia	Ambient	27.8 ± 1.2	38.9 ± 1.7	21.8 ± 0.7	42.7 ± 1.0	15.3 ± 0.9	41.7 ± 2.4	35 ± 1
	Weeks 1–2	20.1 ± 1.0	34.3 ± 1.5	20.7 ± 0.6	42.9 ± 1.4	13.5 ± 0.4	37.6 ± 2.8	42 ± 1
	Weeks 3–4	16.0 ± 0.4	35.8 ± 1.2	20.6 ± 1.0	40.6 ± 0.7	14.1 ± 0.5	34.1 ± 3.0	42 ± 1
	Weeks 5–6	25.7 ± 1.2	34.1 ± 1.4	22.1 ± 1.0	37.5 ± 1.2	11.4 ± 0.8	28.0 ± 2.8	42 ± 1
	Weeks 7–8	26.4 ± 1.3	30.7 ± 2.4	20.7 ± 0.9	37.4 ± 1.3	10.2 ± 0.7	27.2 ± 1.5	33 ± 1
	Continuous	18.8 ± 1.2	23.1 ± 1.1	13.7 ± 0.8	23.1 ± 0.7	3.3 ± 0.2	0.0 ± 0.0	-
	ANOVA	<0.0001	<0.0001	<0.0001	<0.0001	<0.0001	<0.0001	<0.0001
	HSD$_{0.05}$	4.6	6.6	3.5	3.8	2.1	8.1	5
Snapdragon	Ambient	65.2 ± 4.9	60.2 ± 3.1	18.6 ± 0.7	22.5 ± 0.6	7.2 ± 0.7	61.2 ± 8.3	34 ± 1
	Weeks 1–2	66.0 ± 2.0	62.4 ± 3.7	19.5 ± 0.8	23.8 ± 0.6	7.1 ± 0.6	40.3 ± 5.6	41 ± 2
	Weeks 3–4	62.8 ± 3.3	62.2 ± 3.4	20.4 ± 0.4	23.1 ± 0.7	6.2 ± 0.5	57.5 ± 3.5	40 ± 2
	Weeks 5–6	62.7 ± 5.2	73.3 ± 4.0	21.9 ± 1.0	23.4 ± 0.7	6.4 ± 0.6	53.8 ± 5.0	38 ± 2
	Weeks 7–8	63.9 ± 4.6	63.5 ± 3.0	18.8 ± 0.8	22.4 ± 0.6	6.0 ± 0.4	53.3 ± 4.2	34 ± 1
	Continuous	51.2 ± 2.2	51.9 ± 2.5	22.8 ± 1.0	24.1 ± 0.5	4.3 ± 0.3	0.1 ± 0.1	60 ± 2
	ANOVA	0.0199	0.0003	0.0004	0.1822	<0.0001	<0.0001	<0.0001
	HSD$_{0.05}$	13.2	11.9	3.0	-	1.5	16.7	4

[z] Analysis of variance; [y] Tukey's honest significant difference (α = 0.05).

Delayed flowering with the low-energy exposure was primarily due to the lower overall temperature. Increased days to flower in response to decreased temperature have been reported for impatiens 'Accent Red', pansy 'Delta Yellow Blotch', petunia 'Easy Wave Coral Pink' and 'Wave Purple', and snapdragon 'Chimes White', but not for impatiens 'Super Elfin White' [28–32]. For example, a 23 days delay in flowering occurred in snapdragon 'Chimes White' when temperature decreased from 20 to 10 °C [30]. In previous studies, time to flower for petunia 'Easy Wave Coral Pink' and petunia 'Wave Purple' increased as MDT decreased from 26 to 14 °C, and time to flower for pansy 'Delta Yellow Blotch' increased linearly as MDT decreased from 25.7 to 16.3 °C [28,29]. Finally, a 3 to 4 days delay in flowering of impatiens 'Super Elfin Lipstick', petunia 'Avalanche Pink', and pansy 'Colossus Yellow Blotch' was documented for each 1 °C reduction in temperature [33], which is consistent with the 7 to 10 day delay in flowering we observed in our study when providing a two-week low-energy exposure (overall mean temperature was 1.6 to 2.0 °C lower than the ambient treatment; Table 1).

Decreased irradiance in the low-energy conditions may have also contributed to delayed flowering, which has been reported for impatiens 'Super Elfin White', snapdragon 'Rocket Rose', pansy 'Delta Yellow Blotch', and petunia 'Snow Cloud' [29,32,34]. Snapdragon 'Rocket Rose' flowering was delayed 21 days, and impatiens 'Super Elfin White' flowering was delayed 4 days, but only at high temperatures, when DLI decreased from 21.8 to 10.5 $mol \cdot m^{-2} \cdot d^{-1}$ [32]. The influence of DLI on flowering is often attributed to meristem heating by the increased irradiance intensity, and, therefore, is a temperature effect as well [35].

The two-week low-energy exposure also decreased flower number and plant growth relative to those grown at ambient conditions. Flower number in impatiens, pansy, and petunia, but not snapdragon, decreased compared to ambient controls (Table 2). Impatiens was most sensitive, and a two-week low-energy exposure at any point during production reduced flower number 40% to 61%. Pansy flower number was lower only in the weeks 5–6 exposure (35%), and petunia flower number was lower only in the weeks 5–6 and weeks 7–8 exposures (33% and 35%, respectively). Comparing across the four two-week low-energy treatments, timing did not influence impatiens, pansy, or snapdragon flower number, although petunias in low-energy weeks 1–2 had more flowers than the other low-energy intervals. This suggests implementation of a two-week low-energy exposure would not affect flower number regardless of when it was applied during production. Snapdragon inflorescence number was similar across the ambient control and four low-energy treatments (12 to 13 inflorescences), and higher than in the continuous low-energy treatment (<1 inflorescence; data not shown).

Reduced flower number was likely due to both a delay in flowering and reduced net photosynthesis in the low-energy environment. Light is a primary driver of photosynthesis, and temperature influences the rates of enzymatic activity and carbon loss via photorespiration. A decrease in petunia flower development rate as mean DLI decreased from 14 to 4 $mol \cdot m^{-2} \cdot d^{-1}$ has been reported [28]. Likewise, petunia 'Snow Cloud' grown at temperatures ranging from 10 to 30 °C flowered 3 to 23 days later at a given temperature when provided a DLI of 6 $mol \cdot m^{-2} \cdot d^{-1}$ rather than 13 $mol \cdot m^{-2} \cdot d^{-1}$ [34]. A reduction in flower number was observed in petunia 'Supertunia Vista Bubblegum' and 'Supertunia Mini Strawberry Pink Veined' grown 2 days or more per week in low-energy conditions and in pansy 'Matrix Blue Blotch' grown 4 days per week or continuously in low-energy conditions, compared to plants grown continuously in ambient conditions [24].

Relative chlorophyll content index (CCI) four weeks after transplant was generally lower in plants exposed to low-energy conditions, i.e., the weeks 1–2, weeks 3–4, and continuous treatments (Table 2). After eight weeks, relative CCI was not affected by the timing of the low-energy exposure, when compared to ambient conditions. The continuous low-energy treatment had lower CCI values for impatiens and petunia but not pansy or snapdragon (Table 2). This may be related to their cold tolerance, as impatiens and petunia are more cold-sensitive than pansy and snapdragon [25,26]. The reduction in relative CCI is likely a response to the lower temperature. In cotton (*Gossypium hirsutum*

L. var. Delta Pine 61), chlorophyll concentration decreased as temperature decreased [36]. Chlorophyll concentration in tomato (*Solanum lycopersicon* L. cv. M-19) and pepper (*Capsicum annuum* L. cv. M-71) decreased by almost half after a 12 days chilling (5 °C) treatment [37]. Although relative chlorophyll content rather than total chlorophyll concentration was measured in this study, it is an accepted proxy [38].

Plant height was unaffected by the timing of the low-energy exposure in impatiens and petunia. Except for the weeks 7–8 low-energy exposure, which was similar to the ambient control, snapdragon plant height increased with a two-week low-energy exposure (5% to 18% increase; Table 2). We observed an increase in shoot height previously in dianthus 'Telstar Pink' as the number of days per week in low-energy conditions increased [24]. Additionally, shoot height of some *Kalanchoe* species increased as DLI decreased from 17.2 to 4.3 to mol·m^{-2}·d^{-1} [39]. It appears a period of lower DLI during the vegetative phase elicited a shade avoidance response in snapdragon. The timing of the weeks 7–8 low-energy treatment occurred after flowering had begun, and therefore, did not affect internode length nor cause elongation of the inflorescence peduncle.

Plant diameter generally was unaffected or smaller when the timing of the two-week low-energy exposure occurred later in production. Snapdragon plant diameter was similar across all treatments and pansy was smaller only when grown in continuous low-energy conditions (Table 2). Impatiens and petunia plant diameters were smaller than the ambient controls when the two-week low-energy interval occurred later in production, in weeks 3–4 or later or in weeks 5–6 or later, respectively.

Similar to plant height and width, pansy and snapdragon dry mass were smaller only in plants grown continuously in the low-energy conditions (46% and 40%, respectively; Table 2). Petunia and impatiens dry mass gradually decreased as the two-week low-energy exposure occurred later in production. Petunias grown in weeks 5–6 or weeks 7–8 were 26% and 33% smaller, respectively, than ambient controls. Any two-week low-energy exposure in impatiens, regardless of the timing, reduced final dry mass (20% to 31%). The sensitivity of each species to dry mass accumulation corresponds to their cold tolerance; pansy and snapdragon are considered cold-tolerant annuals, petunia is a cold-intermediate species, and impatiens is a cold-sensitive species. The smaller plant size and dry mass is likely the result of reduced photosynthesis at lower irradiance and temperature, which reduced carbohydrate availability for growth. Previously, a decrease in plant dry mass of eight annual bedding plants was observed as DLI decreased, including impatiens 'Cajun Red' and petunia 'Apple Blossom' [40]. Additionally, pansy 'Delta Yellow Blotch' plant mass decreased in response to decreased irradiance, and pansy 'Universal Violet' dry mass decreased as temperature decreased [29,41].

Estimated energy costs were calculated for two production durations. First, energy costs were calculated for the total eight-week production period and relativized to energy costs in the ambient environment. Energy costs in our greenhouse were 84% to 86% of ambient when the two-week low-energy exposure was provided (Table 3). Energy costs increased slightly as the timing of the low-energy period occurred later in production due to naturally increasing temperatures and irradiance from late winter to early spring. The continuous low-energy environment had the lowest energy costs, at 44% of ambient. The second energy calculation estimated energy costs from transplant to mean date of the first flower for each species. This accounted for the lower per day cost in the low-energy conditions but the extended time of production due to delayed flowering. More modest reductions in energy costs occurred, ranging from a 7% reduction to a 17% increase in energy costs compared to the ambient environment (Table 3). Energy costs for the continuous low-energy treatment were not calculated due to the lack of 100% flowering after eight weeks. The estimated energy savings of 4% to 6% when pansy, petunia, and snapdragon were grown in cool, low light conditions during weeks 1–2 of production are slightly lower than the 8% to 18% energy savings reported for other dynamic temperate integration strategies [11,12,16], but much less than the 19% to 46% savings reported [14]. Additionally, snapdragons grown for two weeks in cool, low light conditions generally exhibited the lowest relative energy costs of the four species evaluated, compared to plants grown in ambient conditions, while impatiens generally had the highest relative energy costs. This corresponds to the plant growth and flowering data and is reflective of their classification as cold-tolerant and cold-sensitive species, respectively.

Table 3. Percent energy costs, relative to ambient conditions, calculated in Virtual Grower 3.0.9 using parameters specified [24]. Ambient conditions were 22/18 °C air temperature, 14 h photoperiod, and ambient irradiance + 75 μmol·m^{-2}·s^{-1} supplemental lighting from high-pressure sodium (HPS) lamps when photosynthetic photon flux density (PPFD) was less than 300 μmol·m^{-2}·s^{-1}. Low-energy conditions were 13/10 °C air temperature, ambient irradiance, and a 14 h photoperiod achieved by providing 75 μmol·m^{-2}·s^{-1} from HPS lamps when PPFD was less than 10 μmol·m^{-2}·s^{-1}. Impatiens (*Impatiens walleriana* 'Accent Orange'), pansy (*Viola* × *wittrockiana* 'Delta Premium Blue Blotch'), petunia (*Petunia* × *hybrida* 'Dreams Pink'), and snapdragon (*Antirrhinum majus* 'Montego Violet') were grown in ambient conditions (ambient), grown in ambient conditions and transferred to a low-energy environment at two-week intervals during an eight-week production cycle, or grown continuously in the low-energy environment (continuous).

Treatment	Relative Cumulative Energy Cost [z]	Relative Cumulative Energy Cost at Flowering [y]			
		Impatiens	Pansy	Petunia	Snapdragon
Ambient	100%	100%	100%	100%	100%
Weeks 1–2 cool	84%	103%	96%	94%	95%
Weeks 3–4 cool	86%	101%	98%	98%	93%
Weeks 5–6 cool	86%	99%	102%	99%	97%
Weeks 7–8 cool	88%	117%	106%	94%	100%
Continuous	44%	- [x]	-	-	-

[z] Total energy costs after 8 weeks, reported as a relative percent compared to plants grown continuously in ambient conditions; [y] Energy costs were calculated from transplant to flowering and reported as a relative percent compared to plants grown continuously in ambient conditions; [x] Not all plants had flowered after 8 weeks, and therefore energy costs were not calculated.

One of our objectives was to evaluate the effectiveness of a two-week low-energy interval as a strategy to reduce energy costs for greenhouse production of spring bedding plants. In all species, flowering was the most sensitive and was delayed in all species if the low-energy exposure was applied before flowering. The timing of the two-week low energy exposure influenced the severity of the impact on plant growth. It minimally affected pansy and snapdragon, regardless of when the two-week interval was applied. Petunia was more impacted when the interval was applied in the second half of production, and impatiens dry mass was affected regardless of the week. Therefore, depending on species and target plant size, the timing of the low-energy exposure could be shifted earlier in production to have a minimal impact on plant size and dry mass, or later in production to have a more pronounced effect on final plant size. For example, shifting the exposure later in production may provide a non-chemical strategy for growth regulation of species and cultivars with a vigorous plant habit.

Additionally, we wanted to evaluate the possibility of applying the intermittent short-term reduction in temperature and irradiance as a continuous two-week exposure rather than for 1 to 2 days per week, as reported previously [24], to better align with weather patterns and energy demands. Overall mean temperature and DLI for a crop will be similar when the cumulative duration of low-energy conditions is the same, whether provided continuously for a period of time or intermittently throughout production. Therefore, it stands to reason flowering time would be similar if crops are able to integrate temperature throughout the period of flower initiation and development, as long as temperatures remain between T_{min} and T_{opt}. In our study, pansy, petunia, and snapdragon grown in low-energy conditions for 2 days per week flowered within 2 days of plants exposed to low-energy conditions for a two-week period in weeks 1–2, weeks 3–4, or weeks 4–6 (data not shown). A comparison was not made with impatiens due to insufficient seedling number for the intermittent low-energy treatments.

In conclusion, flowering and plant growth were negatively impacted by the addition of a two-week low-energy exposure, resulting in delayed flowering, fewer flowers, and reduced plant size, mass, and relative chlorophyll content. Cold-sensitive crops, like impatiens, will be most severely impacted. Results from this study indicate the inclusion of a two-week low-energy exposure would decrease energy costs over an eight-week period but have a nominal reduction or even increase in overall energy consumption from transplant to flowering due to delays in flowering. Therefore, it could be worthwhile for growers to consider a short-term intermittent reduction in temperature and DLI to reduce heating costs, but only for cold-tolerant species like snapdragon and if an extra week of production can be built into the production schedule.

Author Contributions: Conceptualization, J.K.B.; Investigation, J.K.B.; Methodology, J.K.B. and J.E.A.; Project administration, J.K.B.; Writing – original draft, J.K.B.; Writing – review and editing, J.K.B. and J.E.A.

Funding: This research received no external funding.

Conflicts of Interest: The authors declare no conflict of interest.

Disclaimer: Mention of trade names or commercial products in this publication is solely for the purpose of providing specific information and does not imply recommendation or endorsement by the U.S. Department of Agriculture. USDA is an equal opportunity provider and employer.

References

1. United States Department of Agriculture. 2012 Census of Agriculture: Census of Horticultural Specialties (2014). AC–12–SS–3. 2015. Available online: https://www.nass.usda.gov/Publications/AgCensus/2012/Online_Resources/Census_of_Horticulture_Specialties/HORTIC.pdf (accessed on 19 November 2018).
2. Runkle, E.; Both, A.J. *Greenhouse Energy Conservation Strategies*; Extension Bulletin E–3160; Michigan State University: East Lansing, MI, USA, 2011.
3. Liu, B.; Heins, R.D. Is plant quality related to the ratio of radiant energy to thermal energy? *Acta Hortic.* **1997**, *435*, 171–182. [CrossRef]
4. Dieleman, J.A.; Meinen, E. Interacting effects of temperature integration and light intensity on growth and development of single-stemmed cut rose plants. *Sci. Hortic.* **2007**, *113*, 182–187. [CrossRef]
5. Dennis, J.H.; Lopez, R.G.; Behe, B.K.; Hall, C.R.; Yue, C.; Campbell, B.J. Sustainable production practices by greenhouse and nursery plant growers. *HortScience* **2010**, *45*, 1232–1237. [CrossRef]
6. Hall, T.J.; Dennis, J.H.; Lopez, R.G.; Marshall, M.I. Factors affecting growers' willingness to adopt sustainable floriculture practices. *HortScience* **2009**, *44*, 1346–1351. [CrossRef]
7. Currey, C.J.; Lopez, R.G.; Mattson, N.S. Finishing bedding plants: A comparison of an unheated high tunnel versus a heated greenhouse in two geographic locations. *HortTechnology* **2014**, *24*, 527–534. [CrossRef]
8. Olberg, M.W.; Lopez, R.G. Growth and development of poinsettia (*Euphorbia pulcherrima*) finished under reduced air temperature and bench-top root-zone heating. *Sci. Hortic.* **2016**, *210*, 197–204. [CrossRef]
9. Camberato, D.M.; Lopez, R.G.; Krug, B.A. Development of *Euphorbia pulcherrima* under reduced finish temperatures. *HortScience* **2012**, *47*, 745–750. [CrossRef]
10. Dieleman, J.A.; Meinen, E.; Marcelis, L.F.M.; de Zwart, H.F.; van Henten, E.J. Optimisation of CO_2 and temperature in terms of crop growth and energy use. *Acta Hortic.* **2005**, *691*, 149–154. [CrossRef]
11. Rijsdijk, A.A.; Vogelezang, J.V.M. Temperature integration on a 24-hour base: A more efficient climate control strategy. *Acta Hortic.* **2000**, *519*, 163–169. [CrossRef]
12. Buwalda, F.; Rijsdijk, A.A.; Vogelezang, J.V.M.; Hattendorf, A.; Batta, L.G.G. An energy efficient heating strategy for cut rose production based on crop tolerance to temperature fluctuations. *Acta Hortic.* **1999**, *507*, 117–125. [CrossRef]
13. Elings, A.; de Zwart, H.F.; Janse, J.; Marcelis, L.F.M.; Buwalda, F. Multiple-day temperature settings on the basis of the assimilate balance: A simulation study. *Acta Hortic.* **2006**, *718*, 219–226. [CrossRef]
14. Lund, J.B.; Andreassen, A.; Ottosen, C.-O.; Aaslyng, J.M. Effect of a dynamic climate on energy consumption and production of *Hibiscus rosa-sinensis* L. in greenhouses. *HortScience* **2006**, *41*, 384–388. [CrossRef]
15. Ottosen, C.-O.; Rosenqvist, E.; Aaslyng, J.M.; Jakobsen, L. Dynamic climate control in combination with average temperature control saves energy in ornamentals. *Acta Hortic* **2005**, *691*, 133–140. [CrossRef]
16. Aaslyng, J.M.; Ehler, N.; Karlsen, P.; Rosenqvist, E. IntelliGrow: A component-based climate control system for decreasing greenhouse energy consumption. *Acta Hortic.* **1999**, *507*, 35–41. [CrossRef]

17. Heins, R.D.; Liu, B.; Runkle, E.S. Regulation of crop growth and development based on environmental factors. *Acta Hortic.* **2000**, *511*, 15–24. [CrossRef]

18. Fink, M. Effects of short-term temperature fluctuations on plant growth and conclusions for short-term temperature optimization in greenhouses. *Acta Hortic.* **1993**, *328*, 147–154. [CrossRef]

19. De Koning, A.N.M. Long-term temperature integration of tomato. Growth and development under alternating temperature regimes. *Scientia Hortic.* **1990**, *45*, 117–127. [CrossRef]

20. Liebig, H.-P. Temperature integration by kohlrabi growth. *Acta Hortic.* **1988**, *230*, 371–380. [CrossRef]

21. Dieleman, J.A.; Meinen, E.; Dueck, T.A. Effects of temperature integration on growth and development of roses. *Acta Hortic.* **2005**, *691*, 51–58. [CrossRef]

22. Körner, O.; Challa, H. Temperature integration and process-based humidity control in chrysanthemum. *Comput. Electron. Agric.* **2004**, *43*, 1–21. [CrossRef]

23. Boldt, J.K.; Gesick, E.Y.; Meyer, M.H.; Erwin, J.E. Alternative periodic energy-efficient light and temperature strategies for herbaceous ornamental production. *HortScience* **2011**, *46*, S354.

24. Boldt, J.K. Short-term reductions in irradiance and temperature minimally affect growth and development of five floriculture species. *HortScience* **2018**, *53*, 33–37. [CrossRef]

25. Blanchard, M.G.; Runkle, E.S. Quantifying the thermal flowering rates of eighteen species of annual bedding plants. *Scientia Hortic.* **2011**, *128*, 30–37. [CrossRef]

26. Vaid, T.M.; Runkle, E.S. Developing flowering rate models in response to mean temperature for common annual ornamental crops. *Scientia Hortic.* **2013**, *161*, 15–23. [CrossRef]

27. LeBude, A.V.; Bilderback, T.E. *The Pour-Through Extraction Procedure: A Nutrient Management Tool for Nursery Crops*; AG–717–W; NC State University Coop Extension Publication: Raleigh, NC, USA, 2009.

28. Blanchard, M.G.; Runkle, E.S.; Fisher, P.R. Modeling plant morphology and development of petunia in response to temperature and photosynthetic daily light integral. *Scientia Hortic.* **2011**, *129*, 313–320. [CrossRef]

29. Niu, G.; Heins, R.D.; Cameron, A.C.; Carlson, W.H. Day and night temperatures, daily light integral, and CO2 enrichment affect growth and flower development of pansy (*Viola* × *wittrockiana*). *J. Am. Soc. Hortic. Sci.* **2000**, *125*, 436–441. [CrossRef]

30. Munir, M.; Jamil, M.; Baloch, J.; Khattak, K.R. Growth and flowering of *Antirrhinum majus* L. under varying temperatures. *Int. J. Agric. Biol.* **2004**, *6*, 173–178.

31. Pramuk, L.A.; Runkle, E.S. Modeling growth and development of celosia and impatiens in response to temperature and photosynthetic daily light integral. *J. Ame. Soc. Hortic. Sci.* **2005**, *130*, 813–818. [CrossRef]

32. Warner, R.M.; Erwin, J.E. Prolonged high temperature exposure and daily light integral impact growth and flowering of five herbaceous ornamental species. *J. Am. Soc. Hortic. Sci.* **2005**, *130*, 319–325. [CrossRef]

33. Mattson, N.S.; Erwin, J.E. Temperature affects flower initiation and development rate of *Impatiens*, *Petunia*, and *Viola*. *Acta Hortic.* **2003**, *624*, 191–197. [CrossRef]

34. Kaczperski, M.P.; Carlson, W.H.; Karlsson, M.G. Growth and development of *Petunia* × *hybrida* as a function of temperature and irradiance. *J. Am. Soc. Hortic. Sci.* **1991**, *116*, 232–237. [CrossRef]

35. Mattson, N.S.; Erwin, J.E. The impact of photoperiod and irradiance on flowering of several herbaceous ornamentals. *Sci. Hortic.* **2005**, *104*, 275–292. [CrossRef]

36. Winter, K.; Königer, M. Dry matter production and photosynthetic capacity in *Gossypium hirsutum* L. under conditions of slightly suboptimum leaf temperatures and high levels of irradiance. *Oecologia* **1991**, *87*, 190–197. [CrossRef] [PubMed]

37. Javanmardi, J.; Rahemi, M.; Nasirzadeh, M. Physiological and reproductive responses of tomato and pepper transplants to low-temperature conditioning. *Int. J. Veg. Sci.* **2013**, *19*, 294–310. [CrossRef]

38. Ruiz-Espinoza, F.H.; Murillo-Amador, B.; García-Hernández, J.L.; Fenech-Larios, L.; Rueda-Puente, E.O.; Troyo-Diéguez, E.; Kaya, C.; Beltrán-Morales, A. Field evaluation of the relationship between chlorophyll content in basil leaves and a portable chlorophyll meter (SPAD-502) readings. *J. Plant Nutr.* **2010**, *33*, 423–438. [CrossRef]

39. Currey, C.J.; Erwin, J.E. Photosynthetic daily light integral impacts growth and flowering of several kalanchoe species. *HortTechnology* **2011**, *21*, 98–102. [CrossRef]

40. Faust, J.E.; Holcombe, V.; Rajapakse, N.C.; Layne, D.R. The effect of daily light integral on bedding plant growth and flowering. *HortScience* **2005**, *40*, 645–649.
41. Adams, S.R.; Pearson, S.; Hadley, P. An analysis of the effects of temperature and light integral on the vegetative growth of pansy cv. Universal Violet (*Viola × wittrockiana* Gams.). *Ann. 3ot.* **1997**, *79*, 219–225. [CrossRef]

horticulturae

MDPI

Article

Water-Related Variables for Predicting Yield of Apple under Deficit Irrigation

Riccardo Lo Bianco

Department of Agricultural, Food and Forest Sciences, University of Palermo, Palermo 90128, Italy;
riccardo.lobianco@unipa.it; Tel.: +39-091-238-96097

Received: 14 December 2018; Accepted: 15 January 2019; Published: 16 January 2019

Abstract: Predicting apple yield in relation to tree water use is important for irrigation planning and evaluation. The aim of the present study was to identify measurable variables related to tree water use that could predict final fruit yield of apple trees under different strategies of deficit irrigation. Adult 'Gala' and 'Fuji' apple trees were exposed to conventional irrigation (CI), delivering 100% of crop evapotranspiration; partial root zone drying (PRD), delivering 50% of CI water only on one alternated side of the root-zone; and continuous deficit irrigation (CDI), delivering 50% of CI water on both sides of the root-zone. Integrals of soil (SWD_{int}) and leaf ($LWSD_{int}$) water deficit along with growth and stomatal conductance (Gs_{int}) were calculated across each season and used to estimate total conductance (GS_{tree}) and transpiration (Tr_{tree}) per tree, transpiration efficiency on a fruit (GR_{fruit}/Tr) or tree (GR_{trunk}/Tr) growth basis, and transpiration productivity ($Yield/Tr_{tree}$). 'Fuji' trees had higher $Yield/Tr_{tree}$, but had lower GR_{trunk}/Tr and similar GR_{fruit}/Tr compared to 'Gala' trees. In 'Fuji', CDI reduced yield, trunk growth, leaf hydration, and gas exchange, while in 'Gala', it did not reduce yield and gas exchange. In 'Fuji', a linear combination of GR_{trunk}/Tr, GR_{fruit}/Tr, and Gs_{tree} contributed to predicting yield, with GR_{fruit}/Tr explaining nearly 78% of the model variability. In 'Gala', a linear combination of $LWSD_{int}$ and Gs_{tree} contributed to predicting yield, with Gs_{tree} explaining over 79% of the model variability. These results indicate that measuring tree water status or water use may help predict final apple yields only in those cultivars like 'Gala' that cannot limit dehydration by closing stomates because of carbon starvation. In more vigorous cultivars like 'Fuji', transpiration efficiency based on fruit growth can be a powerful predictor of final yields.

Keywords: leaf water saturation deficit; partial root zone drying; stomatal conductance; transpiration efficiency; transpiration productivity

---

1. Introduction

On cultivated land, it is estimated that environmental stresses significantly limit agricultural production, and global climate changes are constantly increasing such limitations. Environmental stress tolerance is therefore a critical concern for horticulturists if they hope to increase fruit production as population increases. In particular, plant tissue dehydration (drought stress) may cause direct and indirect decreases in fruit quantity and quality. Indeed, drought may also affect photosynthesis and nutrient uptake causing indirect yield reductions.

A great share of the annual precipitation is lost to evapotranspiration (ET, from 70% up to 90% in arid areas) [1]. This fact proves the importance of adequately estimating the ET component of the hydrologic cycle in predicting on-farm irrigation water management and irrigation planning [2], especially if we consider that without ET there is no production [3]. In particular, the transpiration to ET coefficients have been widely used for precise and efficient irrigation management [4].

Crop yield is determined by both available water quantity and plant water use efficiency [5]. At the physiological level, water use efficiency (WUE) can be defined as the ratio between photosynthesis and transpiration, also defined as transpiration efficiency or instantaneous WUE. This WUE is quite difficult to monitor at the whole tree scale, and even more at the orchard level. For horticultural evaluations, WUE can be more easily expressed as fruit yield per unit of irrigation water, or irrigation water productivity (IWP). Recently, many studies have focused on IWP as being directly related to the increase of WUE [6]. In this study, the use of transpiration efficiency or productivity based on the ratios between relative trunk or fruit growth and transpiration rate as well as the ratio between fruit yield and total tree transpiration is proposed.

Several factors can cause variations of WUE in plants, e.g., air humidity, the different carboxylation mechanism of C3 and C4 plants and, in the long period, the losses due to respiration and assimilate partitioning. Indeed, it has been widely demonstrated that it is possible to improve plant carbohydrate distribution towards reproductive structures, such as fruits, by keeping the plants in a state of mild water deficit, in this way controlling the excessive vegetative growth [7]. This concept has represented in the last decades the basis for a long list of trials investigating the outcomes of what was called "regulated deficit irrigation" (RDI) by Chalmers et al. [8] or "controlled deficit irrigation" by English [9] and [10]. As a matter of fact, trials conducted on several crops showed that IWP tends to increase with deficit compared to conventional irrigation [11,12].

Increasing IWP has been more successful in trees than in field crops for several reasons [13]. Fruit quality, for example, strongly affects crop value, but is not associated with biomass production and water use. In addition, tree fruit growth may not be sensitive to water deficit in certain periods and developmental stages [14]. This, in combination with low volume/high frequency irrigation systems, gives the best opportunities to manage fruit trees under controlled water deficit.

A specific deficit irrigation practice that has received particular attention in the last decades and seems to achieve significant water savings with limited information inputs from the grower is partial root zone drying (PRD) [15,16]. With PRD, one half of the root system is cyclically left to dry; roots in drying soil produce chemical signals (abscisic acid, cytokinins, pH changes), which are translocated to the shoots [17] where they induce partial stomatal closure, reduce transpiration, and ultimately increase WUE [15]. Thanks to the well-watered half of the root zone, the effect on plant water potential is minimal [18] and other metabolic and physiological processes associated to water stress are not affected [15,19]. This deficit irrigation technique has produced positive outcomes in a number of fruit species, and in apple, numerous PRD studies have reported significant increases of IWP and even yields similar to those of full irrigated trees [16].

For the reasons above, understanding the transpiration mechanisms of plants and the factors affecting final crop yield under water deficit becomes a priority. Green plants have indeed many structures and control devices, which allow them to function efficiently even in rapidly changing environments. At the leaf level, transpiration is controlled by physiological and structural factors with stomatal aperture and conductance assuming a primary role [20]. Stomatal conductance (g_s) responds to several factors, such as light, CO_2 concentration, vapor pressure deficit, leaf temperature, leaf abscisic acid, and soil water potential. This latter factor influences g_s by a hormonal signal (abscisic acid) originating in the roots, a sort of biological switch when drought occurs [21].

The prediction of apple yield in relation to water requirement or ET is important for irrigation planning and evaluation. Considerable research has led to the development of simple models for predicting mostly yield of field crops from evapotranspiration during the growing season [22–24]. The aim of the present study was to identify measurable variables related to tree water use that could serve for the development of a model to predict final fruit yield of apple trees under deficit irrigation. The same yield predicting variables and models could be useful for fine tuning of deficit irrigation management.

2. Materials and Methods

Data of the present study are further calculations and analysis of measurements reported in Lo Bianco and Francaviglia [25]. The study was conducted in 2008 and 2009 near Caltavuturo (37°49′ N and 850 m a.s.l.), in central Sicily. Plant material consisted of eight-year-old 'Gala' and 'Fuji' apple trees on M.9 rootstock, trained to a central leader, and spaced at 4 m between rows and 1.5 m within rows. Soil type was a sandy clay loam (53.3% sand, 17.6% silt, and 29.1% clay) with pH 7.3 and 1.8% active carbonates, and soil water potential around −17 kPa at field capacity. With the exception of irrigation, all trees received the same cultural practices.

In the field, two nearby rows (one with 36 'Gala' trees, the other with 36 'Fuji' trees) were selected and divided into four blocks, each including three trees per irrigation treatment. Contiguous irrigation treatments on the same row were separated by two buffer trees. In June, three irrigation treatments were imposed: (1) conventional irrigation (CI), delivering 100% of crop evapotranspiration (ET_c); (2) PRD, where trees received 50% of CI water only on one alternated side of the root zone; (3) continuous deficit irrigation (CDI), where trees received 50% of CI water on both sides of the root zone. Wet and dry sides of PRD trees were alternated every 2–3 weeks when soil water potential in the dry side reached values of approximately −100 to −150 kPa.

Weather parameters were monitored with a μMetos weather station (Pessl, Austria) positioned within the experimental plot and used to determine reference evapotranspiration (ET_0) according to the FAO Penman–Monteith method and crop evapotranspiration (Et_c) [26].

Instantaneous vapor pressure deficit (VPD) was calculated from canopy air temperature (in °C) and relative humidity (in %) measured on the same dates and at the same time as stomatal conductance.

Soil water potential was monitored continuously with six Watermark sensors (Irrometer Co., Riverside, CA, USA) directly connected to the weather station. In drip irrigated apple trees, most of the active roots are within the first 60 cm of soil depth. For this reason, Watermark sensors were positioned at a fixed depth of 40 cm and a distance of about 80 cm from emitters and 1 m from the tree trunk in opposite sides of the root-zone. Integrals of soil water deficit (SWD_{int}) across each irrigation season and treatment were calculated as:

$$SWD_{int} = \Sigma_{(1..t)} \,|\, (SWD_i - SWD_{FC}) \,|, \tag{1}$$

where t is the number of days in the irrigation season, SWD_i are average daily measures of soil water potential, and SWD_{FC} is soil water potential at field capacity. SWD_{int} was used as an indication of soil water deficit accumulated in the root-zone of each treatment during the irrigation periods.

Every two weeks during the irrigation period, at mid-morning two mature, sun-exposed leaves per tree were collected and transported in ice to the laboratory for determination of fresh weight (FW), turgid weight (TW) after rehydrating leaves for 24 h at 8 °C in the dark, and dry weight (DW) after drying leaves at 60 °C to constant weight. Leaf relative water content (RWC) was calculated as [(FW − DW)/(TW − DW)] × 100. Leaf water saturation deficit (LWSD) was calculated as 1 − (RWC/100) and integrated across the irrigation period using the equation proposed by García-Tejero et al. [27] and modified from Myers [28]:

$$LWSD_{int} = \Sigma_{(1..t)} \,|\, LWD_{i+1} \times (n_{i+1} - n_i) + \frac{1}{2} (LWD_i - LWD_{i+1}) \times (n_{i+1} - n_i) \,|, \tag{2}$$

where t is the number of sampling days, LWD_i and LWD_{i+1} are leaf water deficit values measured on two consecutive sampling days (i and i + 1), and n_{i+1} and n_i the days corresponding to two serial samplings. This variable is the integral of a proportion (0 to 1) and therefore can be considered unitless.

On the same dates, stomatal conductance (g_s) was measured with an AP4 Delta-T porometer (Delta-T Devices, Cambridge, UK) on two leaves similar to those used for RWC measurements. Stomatal conductance was also integrated across the irrigation period according the following equation:

$$Gs_{int} = \Sigma_{(1..t)} \,|\, g_{i+1} \times (n_{i+1} - n_i) + \frac{1}{2} (g_i - g_{i+1}) \times (n_{i+1} - n_i) \,|, \tag{3}$$

where t is the number of sampling days, g_i and g_{i+1} are leaf stomatal conductance values measured on two consecutive sampling days (i and i + 1), and n_{i+1} and n_i the days corresponding to two serial measurements.

In each year, one fruit per tree was measured bi-weekly in size (height and width) with a digital caliper. Relative seasonal fruit growth was calculated as the total increase in average diameter (mm) divided by the initial diameter of the fruit (mm). Trunk circumference was measured at about 15 cm above the graft union at the beginning and end of the two growing seasons. Trunk cross-section area (TCSA) was derived from trunk circumference and taken as an indicator of apple tree size [29]. Tree growth was calculated as the increase in TCSA divided by the initial TCSA. Total leaf area per tree (LA) was destructively measured on a separate set of trees from the two cultivars and related to TCSA by regression analysis. The function obtained was used to estimate LA from TCSA measurements in the trees in trial.

Integrals of leaf transpiration (Tr) were derived from Gs_{int} and VPD as follows:

$$Tr = Gs_{int} \times (VPD/101.3), \tag{4}$$

Where 101.3 is the barometric pressure in kPa at sea level. Integrals of soil (SWD_{int}) and leaf ($LWSD_{int}$) water deficit along with growth and stomatal conductance (Gs_{int}) were calculated across each season and used to estimate total conductance (GS_{tree}) and transpiration (Tr_{tree}) per tree, transpiration efficiency on a fruit (GR_{fruit}/Tr) or tree (GR_{trunk}/Tr) growth basis, and transpiration productivity (Yield/Tr_{tree}). Transpiration efficiency on a per tree growth basis was estimated by dividing trunk growth by Tr (Gr_{trunk}/Tr), transpiration efficiency on a fruit growth basis was estimated by dividing fruit growth by Tr (Gr_{fruit}/Tr), while transpiration productivity was obtained by dividing yield by Tr_{tree} (Yield/Tr_{tree}). Transpiration productivity is a very useful measure which is more accurate than IWP and more practical than instantaneous WUE as a trait for improving fruit productivity under limited water resources. Total stomatal conductance per tree was estimated from Gs_{int} and LA, while total transpiration per tree was estimated from Tr and LA.

Data were tested for normal distribution and equal variances and analyzed by analysis of variance and regression procedures using Systat and SigmaPlot software (Systat Software Inc., Richmond, CA, USA). Least squares multiple linear regression with a backward stepwise technique was used to find the best set of variables predicting final apple yield. Means were separated by Tukey's multiple comparison test at $P < 0.05$.

3. Results and Discussion

The relationship between TCSA and LA was described by a non-linear polynomial function, as shown in Figure 1. Canopy and root system size have been linearly related to TCSA in apple [29,30]. In this study with eight-year-old apple trees, the non-linear relationship between TCSA and LA can be explained by canopy size constraints imposed by planting density, tree training form, and pruning. In other words, more vigorous trees (e.g., 'Fuji' trees) were pruned more heavily than weaker trees to remain within the allotted space and avoid competition for light. In this way, canopy and leaf area of trees with different TCSA are brought back to similar sizes determining the observed non-linear relationship between TCSA and LA.

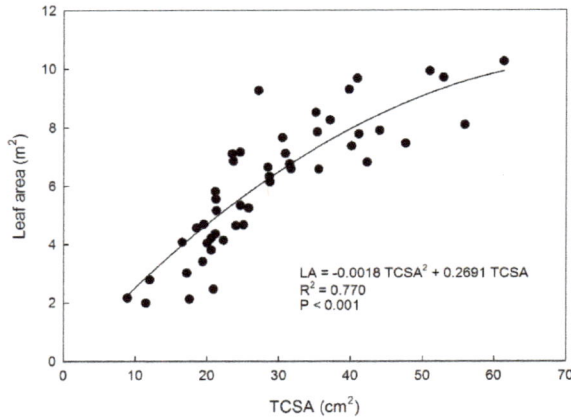

Figure 1. Relationship between trunk cross-section area (TCSA) and total leaf area (LA) in eight-year-old 'Gala' and 'Fuji' apple trees grafted on M.9 rootstock, trained to a central leader, spaced at 4×1.5 m, and grown near Caltavuturo, Sicily.

The imposed irrigation treatments effectively determined the expected differences in soil water deficit. Indeed, on average of the two seasons, SWD_{int} was four times higher in CDI (5.62 MPa) than in CI (1.43 MPa) trees; despite the same irrigation volumes, PRD reported intermediate SWD_{int} (4.61 MPa) with 22% lower values than CDI trees. This has been attributed to greater wetted soil surface and consequent soil evaporation in CDI than in PRD in previous studies [25,31,32].

As expected and regardless of irrigation strategy, 'Fuji' trees were bigger, had greater LA, transpired more water, and yielded more fruit than 'Gala' trees, as shown in Tables 1 and 2. On the other hand, transpiration efficiency in terms of whole tree growth was higher in 'Gala' than in 'Fuji' trees; specifically, 'Gala' trees had higher GR_{trunk}/Tr but similar GR_{fruit}/Tr and lower Yield/Tr_{tree} compared to 'Fuji' trees, as shown in Tables 1 and 2, suggesting that they partitioned assimilates mostly to vegetative rather than fruit growth. This may be at least in part due to their smaller size and fewer constraints to acquire soil resources and fill the allotted space compared to 'Fuji' trees.

The two apple cultivars responded differently to soil water deficit and irrigation strategy. In 'Fuji', PRD maintained plant gas exchange, hydration levels, growth, and productivity similar to CI, while CDI induced significant reductions of yield, tree growth, leaf hydration, and gas exchange, as shown in Table 1. In contrast, no yield and gas exchange differences were observed in deficit irrigated 'Gala' trees, while CDI decreased leaf hydration levels and tree growth compared to both CI and PRD, as shown in Table 2. Transpiration productivity was significantly increased by PRD mainly in 'Fuji' while it was reduced by CDI in 'Gala'. Given the milder soil and leaf water deficit induced by PRD compared to CDI, the different responses of the two cultivars to irrigation may be associated with different levels of water stress resistance, with 'Gala' exhibiting lower ability to limit dehydration than 'Fuji'. In other words, both cultivars tend to close stomates and transpire less as soil water deficit progresses, minimizing symptoms of leaf dehydration (isohydric behavior). Yet, 'Gala' showed higher $LWSD_{int}$ (significant leaf dehydration under both PRD and CDI) than 'Fuji' (leaf dehydration only under CDI). This is at least in part due to the larger 'Fuji' root systems which were able to explore more and deeper soil layers and acquire more water and nutrients than 'Gala' roots.

Table 1. Yield, growth (GR), and water-use related variables in 'Fuji' apple trees under conventional irrigation (CI), partial root zone drying (PRD), and continuous deficit irrigation (CDI).

	CI		PRD		CDI		
Yield (kg/tree)	36.3	ab [z]	39.4	a	33.5	b	* [y]
Yield Efficiency (kg cm^{-2})	0.729	ab	0.853	a	0.607	b	**
GR$_{trunk}$ (cm^2 cm^{-2})	0.051	a	0.052	a	0.037	b	*
GR$_{fruit}$ (mm mm^{-1})	0.541		0.545		0.501		ns
Leaf Area (m^2)	9.02		8.70		9.39		ns
Vapor Pressure Deficit (kPa)	158		159		159		ns
LWSD$_{int}$ [x]	10.1	a	10.5	ab	11.1	b	*
Gs$_{int}$ [w] (mol m^{-2} s^{-1})	16.0	a	15.5	a	12.5	b	**
Tr [v] (mol m^{-2} s^{-1})	25.6	a	24.9	a	20.0	b	**
GR$_{trunk}$/Tr	0.022		0.023		0.019		ns
GR$_{fruit}$/Tr	0.025		0.025		0.028		ns
Yield/Tr$_{tree}$	0.177	b	0.210	a	0.191	ab	*
Tr$_{tree}$ [u] (mol s^{-1}/tree)	229		216		188		ns
Gs$_{tree}$ [u] (mol s^{-1}/tree)	143		136		118		ns

[z]   Mean separation within rows by Tukey's multiple comparison test at $P < 0.05$

[y]   Level of statistical significance for irrigation factor from analysis of variance: ns, $P > 0.05$; *, $P < 0.05$; **, $P < 0.01$; ***, $P < 0.001$.

[x]   LWSD$_{int}$ is leaf water saturation deficit integrated across the irrigation period.

[w]   Gs$_{int}$ is stomatal conductance integrated across the irrigation period.

[v]   Tr is transpiration integrated across the irrigation period.

[u]   Seasonal integrals of total conductance (GS$_{tree}$) and transpiration (Tr$_{tree}$) per tree.

Table 2. Yield, growth (GR), and water-use related variables in 'Gala' apple trees under conventional irrigation (CI), partial root zone drying (PRD), and continuous deficit irrigation (CDI).

	CI		PRD		CDI		
Yield (kg/tree)	19.3		18.5		17.2		ns [z]
Yield$_\text{Eff}$ (kg cm^{-2})	0.600		0.659		0.537		ns
GR$_\text{trunk}$ (cm^2 cm^{-2})	0.082	a [y]	0.071	a	0.053	b	***
GR$_\text{fruit}$ (mm mm^{-1})	0.545		0.568		0.563		ns
Leaf Area (m^2)	6.69		6.31		6.95		ns
Vapor Pressure Deficit (kPa)	169		168		170		ns
LWSD$_\text{int}$ [x]	7.53	a	8.28	ab	8.93	b	*
Gs$_\text{int}$ [w] (mol m^{-2} s^{-1})	13.9		12.9		12.4		ns
Tr [v] (mol m^{-2} s^{-1})	23.7		21.0		22.0		ns
GR$_\text{trunk}$/Tr	0.044	a	0.044	a	0.029	b	*
Gr$_\text{fruit}$/Tr	0.028		0.033		0.028		ns
Yield/Tr$_\text{tree}$	0.141	ab	0.161	a	0.124	b	*
Tr$_\text{tree}$ [u] (mol s^{-1}/tree)	154		131		150		ns
Gs$_\text{tree}$ [u] (mol s^{-1}/tree)	90.5		77.4		88.5		ns

[z] Level of statistical significance for irrigation factor from analysis of variance: ns, $P > 0.05$; *, $P < 0.05$; **, $P < 0.01$; ***, $P < 0.001$.
[y] Mean separation within rows by Tukey's multiple comparison test at $P < 0.05$
[x] LWSD$_\text{int}$ is leaf water saturation deficit integrated across the irrigation period.
[w] Gs$_\text{int}$ is stomatal conductance integrated across the irrigation period.
[v] Tr is transpiration integrated across the irrigation period.
[u] Seasonal integrals of total conductance (GS$_\text{tree}$) and transpiration (Tr$_\text{tree}$) per tree.

In 'Fuji', a linear combination of GR$_\text{trunk}$/Tr, GR$_\text{fruit}$/Tr, and Gs$_\text{tree}$ contributed to predicting yield of apple trees under soil water deficit, as shown in Table 3. In this cultivar, GR$_\text{fruit}$/Tr was the most important variable for predicting yield, explaining nearly 78% of the model variability, while GR$_\text{trunk}$/Tr can be considered negligible as it explained only about 3% of the model variability, as shown in Table 3. On the other hand, a linear combination of LWSD$_\text{int}$ and Gs$_\text{tree}$ contributed to predicting yield of 'Gala' apple trees under soil water deficit, as shown in Table 4. In this cultivar, Gs$_\text{tree}$ was the most important variable for predicting yield, explaining over 79% of the model variability, as shown in Table 4. This difference between the two cultivars indicates that fruit yield of 'Gala' trees was more sensitive to stomatal closure compared to 'Fuji' trees. This may be due to differences in tree size and leaf water deficit levels. The increase of LWSD$_\text{int}$ over the control was indeed greater in 'Gala' (0.75 and 1.4 for PRD and CDI, respectively) than in 'Fuji' (0.4 and 1 for PRD and CDI, respectively). The effect of tree water status on apple yield has been already documented, although other factors like crop load may have stronger effects than water deficit on yield [33]. In addition, the larger 'Fuji' trees may have been less sensitive to stomatal closure than 'Gala' trees because of greater carbon and water storage in permanent structures. In this regard, others have reported contrasting results indicating positive or no effect of tree size and capacitance on water status [34,35], while there is little doubt about the role of permanent structures as carbon reservoirs. In addition to carbon reserves in permanent structures, differences in LA and photosynthetic rates as well as nutrient acquisition may play a significant role in the response of the two cultivars. In the present study, differences in tree size and carbon and water storage may also help explain the higher transpiration productivity in 'Fuji' than in 'Gala' and the major contribution of Gr$_\text{fruit}$/Tr to yield prediction in 'Fuji'. LWSD$_\text{int}$ was a relatively weak yield predictor only in 'Gala', suggesting that even under soil water limiting conditions, factors other than water (e.g., assimilation rate, nutrient status, flower fertility, pollination) are major determinants of apple fruit yield formation and a simple measurement of tree water status may not serve as a solid yield predictor.

Table 3. Multiple linear regression model and parameters contributing to predict yield in 'Fuji' apple trees under deficit irrigation.

Yield = 28.5 + (2774 × GR_{trunk}/Tr) − (295 × GR_{fruit}/Tr) + (0.070 × Gs_{tree})					
N = 65	R = 0.554	R^2 = 0.306	SE of Estimate = 8.85		F < 0.001
Parameters	Coefficient	SE	t	P	% of SSreg
Constant	28.5	6.74	4.23	<0.001	-
GR_{trunk}/Tr [z]	2774	973	2.85	0.006	3.1
GR_{fruit}/Tr [y]	−295	118	−2.51	0.015	77.9
Gs_{tree} [x]	0.070	0.031	2.26	0.027	19.0

[z]   Trunk growth/transpiration.
[y]   Fruit growth/transpiration.
[x]   Seasonal integral of total conductance (Gs) per tree.

Table 4. Multiple linear regression model and parameters contributing to predict yield in 'Gala' apple trees under deficit irrigation.

Yield = −0.019 + (1.35 × $LWSD_{int}$) + (0.085 × Gs_{tree})					
N = 68	R = 0.410	R^2 = 0.168	SE of Estimate = 6.33		P = 0.003
Parameters	Coefficient	SE	t	P	% of SSreg
Constant	−0.019	5.50	−0.003	0.997	-
$LWSD_{int}$ [z]	1.35	0.52	2.59	0.012	20.5
Gs_{tree} [y]	0.085	0.026	3.23	0.002	79.5

[z] $LWSD_{int}$ is leaf water saturation deficit integrated across the irrigation period.
[y]   Seasonal integral of total conductance (Gs) per tree.

In conclusion, the more vigorous 'Fuji' trees were more efficient than 'Gala' trees under soil water deficits in terms of yield and transpiration productivity. Our results indicate that measuring tree water status or gas exchange may help predict final apple yields only in those trees and cultivars (like 'Gala' in this study) that are not able to limit dehydration by closing stomates because of carbon starvation. In more vigorous trees and cultivars like 'Fuji', transpiration (or water use) efficiency towards fruit growth seems to be a powerful predictor of final yields.

Funding: This research received no external funding.

Conflicts of Interest: The author declares no conflict of interest.

References

1.   Hamon, R.W. Evapotranspiration and water yield predictions. In Evapotranspiration and its role in water resources management. *Conf. Proc. Am. Soc. Agric. Eng.* **1966**, *December*, 8–9.
2.   Shockley, D.G. Evapotranspiration and farm irrigation planning and management. In Evapotranspiration and its role in water resources management. *Conf. Proc. Am. Soc. Agric. Eng.* **1966**, *December*, 3–5.
3.   Hansen, V.E. Evapotranspiration and water resources management. *Conf. Proc. Am. Soc. Agric. Eng.* **1966**, *December*, 12–13.
4.   Kang, S.; Gu, B.; Du, T.; Zhang, J. Crop coefficient and ratio of transpiration to evapotranspiration of winter wheat and maize in a semi-humid region. *Agric. Water Manag.* **2003**, *59*, 239–254. [CrossRef]

5. Xu, L.K.; Hsiao, T.C. Predicted versus measured photosynthetic water-use efficiency of crop stands under dynamically changing field environments. *J. Exp. Bot.* **2004**, *55*, 2395–2411. [CrossRef]

6. Kijne, J.W.; Barker, R.; Molden, D. *Water Productivity in Agriculture: Limits and Opportunities for Improvement*; CABI Publishing: Wallingford, UK, 2003; 332p.

7. Chalmers, D.J.; Mitchell, P.D.; Vanheek, L. Control of peach tree growth and productivity by regulated water supply, tree density and summer pruning. *J. Am. Soc. Hortic. Sci.* **1981**, *106*, 307–397.

8. Chalmers, D.J.; Burge, G.; Jerie, P.H.; Mitchell, P.D. The mechanism of regulation of Bartlett pear fruit and vegetative growth by irrigation withholding and regulated deficit irrigation. *J. Am. Soc. Hortic. Sci.* **1986**, *111*, 904–907.

9. English, M.J. Deficit irrigation: Analytical framework. *J. Irrig. Drain. Eng.* **1990**, *116*, 399–412. [CrossRef]

10. Mitchell, P.D.; Jerie, P.H.; Chalmers, D.J. Effects of regulated water deficits on pear tree growth, flowering, fruit growth and yields. *J. Am. Soc. Hortic. Sci.* **1984**, *109*, 604–606.

11. Zwart, S.J.; Bastiaanssen, W.G. Review of measured crop water productivity values for irrigated wheat, rice, cotton and maize. *Agric. Water Manag.* **2004**, *69*, 115–133. [CrossRef]

12. Fan, T.; Wang, S.; Xiaoming, T.; Luo, J.; Stewart, B.A.; Gao, Y. Grain yield and water use in a long-term fertilization trial in Northwest China. *Agric. Water Manag.* **2005**, *76*, 36–52. [CrossRef]

13. Fereres, E.; Goldhamer, D.A.; Parsons, L.R. Irrigation water management of horticultural crops. *HortScience* **2003**, *38*, 1036–1042.

14. Johnson, R.S.; Handley, D.F. Using water stress to control vegetative growth and productivity of temperate fruit trees. *HortScience* **2000**, *35*, 1048–1050.

15. Dry, P.R.; Loveys, B.R.; Botting, D.G.; Düring, H. Effects of partial root-zone drying on grapevine vigour, yield, composition of fruit and use of water. In Proceedings of the Ninth Australian Wine Industry Technical Conference, Adelaide, South Australia, 16–19 July 1995; pp. 128–131.

16. Lo Bianco, R. Responses of apple to partial root-zone drying. A review. In *Irrigation Management, Technologies, and Environmental Impacts*; Ali, M.H., Ed.; Nova Science Publishers, Inc.: New York, NY, USA, 2013; Chapter 3; pp. 71–86.

17. Dodd, I.C.; Egea, G.; Davies, W.J. Abscisic acid signalling when soil moisture is heterogeneous: Decreased photoperiod sap flow from drying roots limits ABA export to the shoots. *Plant Cell Environ.* **2008**, *31*, 1263–1274. [CrossRef] [PubMed]

18. Gowing, D.J.G.; Davies, W.J.; Jones, H.G. A positive root-sourced signal as an indicator of soil drying in apple, *Malus* × *domestica* Borkh. *J. Exp. Bot.* **1990**, *41*, 1535–1540. [CrossRef]

19. Dry, P.R.; Loveys, B.R.; Düring, H. Partial drying of the root-zone of grape. I. Transient changes in shoot growth and gas exchange. *Vitis* **2000**, *39*, 3–7.

20. Cowan, I.R.; Farquhar, G.D. Stomatal function in relation to leaf metabolism and environment. *Symp. Soc. Exp. Biol.* **1977**, *31*, 471. [PubMed]

21. Davies, W.J.; Zhang, J. Root signals and the regulation of growth and development of plants in drying soil. *Ann. Rev. Plant Biol.* **1991**, *42*, 55–76. [CrossRef]

22. Doorenbos, J.; Kassam, A.H. Yield response to water. In *Irrigation and Drainage Paper No. 33*; FAO: Rome, Italy, 1986.

23. Howel, T.A.; Musick, J.T. Relationship of dry matter production of field crops to water consumption. In *Crop Water Requirements*; Perrier, A., Riou, C., Eds.; INRA: Paris, France, 1985; pp. 247–269.

24. Ouda, S.A.; Khalil, F.A.; Tantawy, M.M. Predicting the impact of water stress on the yield of different maize hybrids. *Res. J. Agric. Biol. Sci.* **2006**, *2*, 369–374.

25. Lo Bianco, R.; Francaviglia, D. Comparative responses of 'Gala' and 'Fuji' apple trees to deficit irrigation: Placement versus volume effects. *Plant Soil* **2012**, *357*, 41–58. [CrossRef]

26. Allen, R.G.; Pereira, L.S.; Raes, D.; Smith, M. Crop evapotranspiration—Guidelines for computing crop water requirements. In *Irrigation and Drainage Paper 56*; FAO: Rome, Italy, 1998.

27. García-Tejero, I.; Romero-Vicente, R.; Jiménez-Bocanegra, J.A.; Martínez-García, G.; Durán-Zuazo, V.H.; Muriel-Fernández, J.L. Response of citrus trees to deficit irrigation during different phenological periods in relation to yield, fruit quality, and water productivity. *Agric. Water Manag.* **2010**, *97*, 689–699. [CrossRef]

28. Myers, B.J. Water stress integral—A link between short term stress and long term growth. *Tree Physiol.* **1988**, *4*, 315–323. [CrossRef]

29. Lo Bianco, R.; Policarpo, M.; Scariano, L. Effect of rootstock vigor and in-row spacing on stem and root growth, conformation, and dry matter distribution of young apple trees. *J. Hortic. Sci. Biotechnol.* **2003**, *78*, 828–836. [CrossRef]

30. Westwood, M.N.; Roberts, A.N. The relationship between trunk cross-sectional area and weight of apple trees. *J. Am. Soc. Hortic. Sci.* **1970**, *95*, 28–30.

31. Marsal, J.; Mata, M.; Del Campo, J.; Arbones, A.; Vallverdú, X.; Girona, J.; Olivo, N. Evaluation of partial root-zone drying for potential field use as a deficit irrigation technique in commercial vineyards according to two different pipeline layouts. *Irrig. Sci.* **2008**, *26*, 347–356. [CrossRef]

32. Mossad, A.; Scalisi, A.; Lo Bianco, R. Growth and water relations of field-grown 'Valencia' orange trees under long-term partial rootzone drying. *Irrig. Sci.* **2018**, *36*, 9–24. [CrossRef]

33. Naor, A.; Naschitz, S.; Peres, M.; Gal, Y. Responses of apple fruit size to tree water status and crop load. *Tree Physiol.* **2008**, *28*, 1255–1261. [CrossRef]

34. Davies, F.S.; Lakso, A.N. Diurnal and seasonal changes in leaf water potential components and elastic properties in response to water stress in apple trees. *Physiol. Plant.* **1979**, *46*, 109–114. [CrossRef]

35. Olien, W.C.; Lakso, A.N. Effect of rootstock on apple (*Malus domestica*) tree water relations. *Physiol. Plant.* **1986**, *67*, 421–430. [CrossRef]

horticulturae

MDPI

Article

How Water Quality and Quantity Affect Pepper Yield and Postharvest Quality

Elazar Fallik [1,*], Sharon Alkalai-Tuvia [1], Daniel Chalupowicz [1], Merav Zaaroor-Presman [1], Rivka Offenbach [2], Shabtai Cohen [2] and Effi Tripler [2]

[1] Agricultural Research Organization, The Volcani Center, Department of Postharvest Science of Fresh Produce, Rishon Leziyyon 7505101, Israel; sharon@volcani.agri.gov.il (S.A.-T.); chalu@volcani.agri.gov.il (D.C.); merav.zaaroor@mail.huji.ac.il (M.Z.-P.)
[2] Central and Northern Arava Research and Development, Arava Sapir 8682500, Israel; Rivka@arava.co.il (R.O.); sab@inter.net.il (S.C.); effi@arava.co.il (E.T.)
* Correspondence: efallik@volcani.agri.gov.il; Tel.: +972-3-9683665

Received: 29 November 2018; Accepted: 2 January 2019; Published: 7 January 2019

Abstract: There are gaps in our knowledge of the effects of irrigation water quality and amount on yield and postharvest quality of pepper fruit (*Capsicum annuum* L.). We studied the effects of water quality and quantity treatments on pepper fruits during subsequent simulated storage and shelf-life. Total yield decreased with increasing water salinity, but export-quality yield was not significantly different in fruits irrigated with water of either 1.6 or 2.8 dS/m, but there was a 30–35% reduction in export-quality yield following use of water at 4.5 dS/m. Water quantity hardly affected either total or export-quality yield. Water quality but not quantity significantly affected fruit weight loss after 14 days at 7 °C plus three days at 20 °C; irrigation with water at 2.8 dS/m gave the least weight loss. Fruits were significantly firmer after irrigation with good-quality water than with salty water. The saltier the water, the higher was the sugar content. Vitamin C content was not affected by water quality or quantity, but water quality significantly affected antioxidant (AOX) content. The highest AOX activity was found with commercial quality water, the lowest with salty water. Pepper yield benefited by irrigation with fresh water (1.6 dS/m) and was not affected by water quantity, but post-storage fruit quality was maintained better after use of moderately-saline water (2.8 dS/m). Thus, irrigation water with salinity not exceeding 2.8 dS/m will not impair postharvest quality, although the yield will be reduced at this salinity level.

Keywords: prolonged storage; salinity; shelf-life

1. Introduction

The amount of agricultural land destroyed by salt accumulation each year, worldwide, is estimated to be 10 million ha [1]. Furthermore, this destruction rate could be accelerated by: climate change; excessive use of groundwater; increasing use of low-quality water in irrigation; and the massive introduction of irrigation associated with intensive farming. On the other hand, it has been confirmed in many regions that the tendency to increase the efficiency of irrigation water use and to irrigate with low-quality water, because of water scarcity, can lead to accumulation of salts in the soil. It is estimated that by 2050, 50% of the world's arable land will be affected by salinity [2].

During the last decade, salinity and drought were two of the major abiotic stresses in the Arava Valley in the southern part of Israel. This region is predominantly arid and is affected by salinity because of very low rainfall (<30 mm year^{-1}), high evapotranspiration (3000 mm year^{-1}), and groundwater that is mostly saline, with an electrical conductivity (EC) about 2.8 dS/m. Moreover, the amount of water available for irrigation is declining every year; salinity is gradually increasing and there are underground water wells with more than 4 dS/m. Consequently, plant growth and yield can be negatively affected [3]. Azuma et al. [4] reported that the detrimental impact of salinity mainly affects fruits rather than leaves and stems. Thus, high salinity and water scarcity in agricultural soils present the most serious challenges faced by horticultural crops in southern Israel.

The major crop in the Arava Valley during the winter is sweet bell pepper (*Capsicum annuum* L.); about 60% of the sweet bell pepper that is designated for export from Israel is grown in this region during the fall and winter; the growth area is estimated at 2000 ha, with an average yield of about 80–120 ton ha^{-1}. Pepper plants are sensitive to drought stress and moderately sensitive to salt stress [5,6]. Nevertheless, very little is known about the influence of water quantity and quality on pepper fruit quality after harvest and prolonged storage. Therefore, the objective of the present study was to evaluate, for two consecutive years, the effects of water quantity (i.e., irrigation water), and quality (i.e., salinity) on pepper yield and fruit quality after prolonged storage and shelf-life simulation.

2. Materials and Methods

2.1. Plant Materials and Physical Design

The study was performed at Yair experimental station (30°46′45.3″ N; 35°14′31.1″ E) in a 900 m^2 greenhouse, situated in Israel's Central Arava Valley, 130 m below mean sea level. The experiment took place during the growing season 2015/2016, in which sweet red bell-pepper (*Capsicum annuum* L., cv. Cannon) was evaluated for yield, fruit quality and postharvest indicators. The local soil texture is loamy sand, having sand, silt and clay percentages of 83, 8 and 9%, respectively [7]. Two row crops of pepper seedlings were planted on 5 August 2015, in each bed and spaced 0.4 × 0.4 m. The distance between each bed was 1.6 m, which yielded a planting density of 31,250 plants·ha^{-1}. The experiments were equipped with a pressure-compensated drip irrigation system (Netafim Ltd., Hatzerim, Israel), consisting of one lateral for each crop row having an outer diameter of 0.017 m. The integrated drippers were spaced 0.2 m and their discharge was 1.6 L h^{-1}.

Prior to the planting, the greenhouse was enclosed with 25 mesh insect net, with an additional net-shading on the roof which reduced the radiation by 30%. The net shade was removed 6 weeks after the planting, followed by enclosing the greenhouse with translucent plastic (0.12 mm thick, IR—Ginegar Plastic Ltd., Kibbutz Ginegar, Israel), 1 month later. A Spanish trellising method was applied and common cultivation (leaf pruning, side shoots removal, vine-training and canopy-height adjustment) and plant protection practices were used throughout the growing season [8]. Temperature measurements records were downloaded from an adjacent Israeli Meteorological Services (IMS) meteorological weather station.

2.2. Irrigation and Yield

The experimental design was randomized blocks (n = 4), with 20 plants in each replicate. Three irrigation water salinities (EC 1, 2.8 and 4 dS·m^{-1}) and 3 water application levels were applied for each water quality (Table 1). Irrigation application levels were determined based on the long-term (2002–2014) averages of potential evapotranspiration rates of bell peppers in the Arava region. Electrical conductivity of 1 dS·m^{-1} was applied by blending local saline water (EC = 2.8 dS·m^{-1} with desalinated water, while the highest salinity level (EC of 4.5 dS·m^{-1}) was achieved by an equivalent addition of sodium chloride and calcium chloride salts to the local saline water.

Table 1. Irrigation water salinities and their specific application levels, since Day After Planting (DAP). Fertilizer solution in irrigation water contained N as total nitrogen, P as P_2O_5 and K as K_2O.

		Electrical Conductivity of the Irrigation Water (dS·m^{-1})										
		1			2.8			4				
		Water Application Levels (% from ET_p)									Fertilizer Application	
DAP	ET_p	70	100	150	100	150	200	100	200	300	N-P-K	N
	(mm·d^{-1})	Daily Irrigation Water Depths (mm·d^{-1})									(%)	(mg·L^{-1})
0–35	1.3	0.91	1.3	1.95	1.3	1.95	2.6	1.3	2.6	3.9	6-6-6	50
36–51	3.3	2.31	3.3	4.95	3.3	4.95	6.6	3.3	6.6	9.9	6-6-6	50
52–62	2.7	1.89	2.7	4.05	2.7	4.05	5.4	2.7	5.4	8.1	7-3-7	120
63–94	2.5	1.75	2.5	3.75	2.5	3.75	5	2.5	5	7.5	7-3-7	150
95–104	1.7	1.19	1.7	2.55	1.7	2.55	3.4	1.7	3.4	5.1	7-3-7	100
105–114	1.2	0.84	1.2	1.8	1.2	1.8	2.4	1.2	2.4	3.6	7-3-7	100
115–124	1.2	0.84	1.2	1.8	1.2	1.8	2.4	1.2	2.4	3.6	7-3-7	100
125–134	0.8	0.56	0.8	1.2	0.8	1.2	1.6	0.8	1.6	2.4	4-2-6	100
135–144	0.8	0.56	0.8	1.2	0.8	1.2	1.6	0.8	1.6	2.4	4-2-6	100
145–154	1.1	0.77	1.1	1.65	1.1	1.65	2.2	1.1	2.2	3.3	4-2-6	100
155–164	1.3	0.91	1.3	1.95	1.3	1.95	2.6	1.3	2.6	3.9	4-2-6	100
165–194	2	1.4	2	3	2	3	4	2	4	6	4-2-6	100
195–224	3	2.1	3	4.5	3	4.5	6	3	6	9	4-2-6	100
225–243	4	2.8	4	6	4	6	8	4	8	12	4-2-6	100
244–272	5	3.5	5	7.5	5	7.5	10	5	10	15	4-2-6	100

Yield data included the cumulative weight of fruits (total yield) and that of defect-free fruits (export-quality yield), from December through mid-March from all four repetitions per treatment. Results are expressed in ton ha^{-1}.

Petioles were sampled from newly fully-expanded leaves located at the 4th petiole from the apex. Approximately 20–25 petioles were collected at random from each replicate. The samples were taken between 8:00–10:00 am to minimize differences in cell turgidity of plants. The leaflets were stripped, and the petioles placed in a zip-lock bag. One mL of freshly pressed sap was diluted with 50 mL of distilled water. The solution was analyzed for chloride concentration by means of a standard chloridometer instrument.

2.3. Postharvest Fruit Quality Parameters

The postharvest quality was determined once monthly, at the end of December, at the beginning of February, and mid-March of each year; there were three harvests per year. Each harvest was collected in four corrugated cartons, each containing 5 kg of export-quality pepper fruits. The fruits were of uniform size of 180–200 g, at 85–90% maturity, with attached calyx and free of defects. Immediately after harvest, fruits were rinsed and brushed in hot water as described by Fallik et al. [9]. Fruit-quality parameters were evaluated immediately after each harvest and at the end of 14 days of storage at 7 °C and relative humidity (RH) of ~95%, followed by 3 days at 20 °C. Weight loss was expressed as percentage loss from the initial weight of 10 fruits. Fruit flexibility was measured by placing the fruit between two horizontal flat plates, the upper of which was loaded with a 2-kg weight, as described by Fallik et al. [9]. A dial fixed to a graduated plate recorded the deformation of the fruit in millimeters. Full deformation was measured 15 s after placing the load on the fruit, the weight was removed, and the residual deformation was measured after a further 15 s. The residual deformation directly indicated fruit elasticity: a fruit with 0–1.5 mm deformation was designated as very firm; with 1.6–3.0 mm deformation as firm; with 3.1–4.5 mm deformation as soft; and with more than 4.6 mm deformation as very soft. Total soluble solids (TSS) were measured in the five fruits that had been tested for firmness, by squeezing juice out of the fruits and recording the readings on an Atago digital refractometer

(Atago, Tokyo, Japan). A fruit was considered decayed if fungal mycelia appeared on the peel or calyx, and decay was expressed as the percentage of decayed fruits in the carton.

The vitamin C content of the bell pepper fruits was determined with the HI3850 Ascorbic Acid Test Kit (Hanna Instruments, Smithfield, RI, USA), which expresses measured quantities as milligrams per 100 g. In accordance with the test kit instructions, 2 g of fresh bell pepper fruit was homogenized with 10 mL of deionized water in a 50-mL vial at high speed for 1 min. The homogenate was passed through filter paper and kept on ice pending mixing of a 1-mL aliquot of homogenate with 49 mL deionized water in a beaker. Then 1 mL of HI3850A-0 reagent and four drops of starch as an indicator were added, and HI3850C-0 reagent was added as 10-mL drops, which were counted until a persistent blue color was developed when the beaker was swirled.

Antioxidant activity (AOX) was measured by using the discoloration method [10] based on 2,2′-azinobis (3-ethylbenzothiazoline-6-sulfonate) (ABTS⁺) (Sigma-Aldrich, Rehovot, Israel) with slight modification. In the present study, only hydrophilic fractions were isolated from 100 mg of freeze-dried powder by stepwise extraction with acetate buffer, acetone, and hexane, and repeated partitioning of water-soluble and -insoluble portions. Antioxidant activity was evaluated by discoloration of the ABTS⁺ radical cation. The radical was generated in acetate buffer medium at pH 4.3 to facilitate the activities of the hydrophilic antioxidants. The final reaction mixture contained 150 μmol of ABTS⁺ and 75 μmol of potassium persulfate ($K_2S_2O_8$) in 249 mL of acetate buffer at pH 4.3. Incubation of the reaction mixture at 45 °C for 1 h was sufficient to generate ABTS⁺. The resulting stock solution of ABTS⁺ can be stored for up to 3 days at 4 °C without significant loss of properties. The discoloration test was performed in a 96-well microplate by adding 3 μL of test sample to 300 μL of ABTS⁺ and comparing the optical density at 734 nm after 15 min of incubation at room temperature, with that of a blank sample. Final results were calculated by comparing the absorbance of the samples with that of the standard (±)-6-hydroxy-2,5,7,8-tetramethylchromane-2-carboxylic acid (Trolox) (Sigma-Aldrich). The antioxidant activity in the samples was determined as Trolox equivalents (TE), according to the formula

$$TE = (A_{sample} - A_{blank})/(A_{standard} - A_{blank}) \times C_{standard}$$

where A is the absorbance at 734 nm and C is the concentration of Trolox (mmol).

The TE antioxidant capacity (TEAC) per unit weight of plant tissue was calculated as follows:

$$TEAC \ (mmol \ TE/mg) = (TE \times V)/(1000 \times M)$$

in which V is the final extract volume and M is the amount of tissue extracted.

The contents of vitamin C and antioxidant activity were measured in 10 fruits taken from each treatment, at each of the three harvests each year.

2.4. Statistical Analysis

The data shown here are the means of two consecutive experiments with three harvests each year; the results were similar. The results were subjected to two-way analysis of variance (ANOVA) with JMP 11 version (SAS, Cary, NC, USA). The means were separated by using the Least Significance Difference (LSD) test at $p < 5\%$. Pairwise correlation analysis was carried out to determine the significance level of the correlation between the parameters of interest.

3. Results

3.1. Yield and Chloride in Petiole

The better the water quality, the higher was the total cumulative yield during the growing season (Figure 1); the total yield decreased as the water salinity increased. The average total yield with water of EC 1.6 dS m^{-1} was about 128-ton ha^{-1}; in water quality of 2.8 and 4.5 dS m^{-1} the average total yield was 115- and 99-ton ha^{-1}, respectively. The export-quality yields were not significantly different at both 1.6 and 2.8 dS m^{-1}, with an average yield of 70- and 65-ton ha^{-1}, respectively. Reductions of 35 and 30% in export-quality yield were observed when plants were irrigated with water at 4.5 dS m^{-1}, compared with those obtained at 1.6 and 2.8 dS m^{-1}, respectively (Figure 1). Water quantity hardly affected either total or export-quality yield, although there was a slight increase in total yield with irrigation at 1.6–1.5 dS m^{-1} and a slight decrease in total yield at 4.5–2.0 or 4.5–3.0 dS m^{-1}.

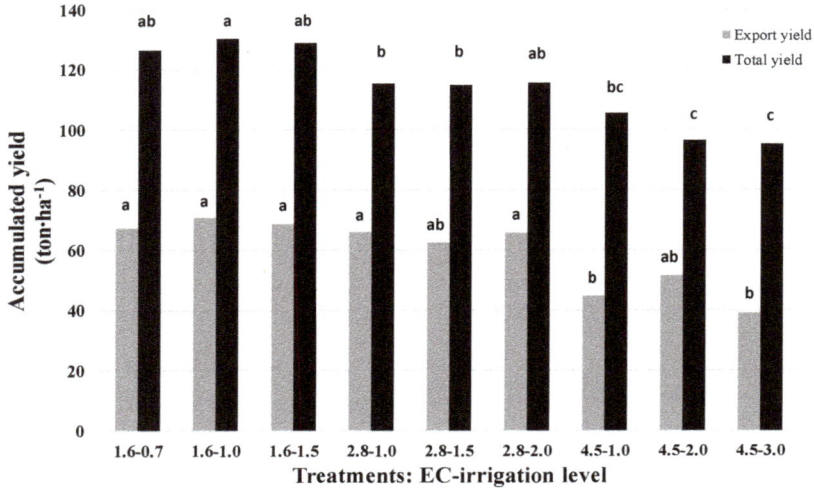

Figure 1. The influence of water quality (electrical conductivity (EC) of the irrigation—1.6, 2.8 and 4.5 dS m^{-1} EC) and water quantity on the cumulative total and export-quality yields of pepper between December and mid-March. Means of columns with the same letter are not significantly different (LSD; $p < 0.05$).

High chloride concentrations were reordered in petioles of peppers treated with low irrigation levels (0.7 and 1). Numerically, at salinities of 1, 2.8 and 4.5 dS·m^{-1}, the chloride concentrations were 150, 159 and 197.5 mg·L^{-1}, respectively. A further increase in irrigation level reduced chloride content in the petioles. However, no differences were observed between the two high water application levels in each salinity treatment (Figure 2).

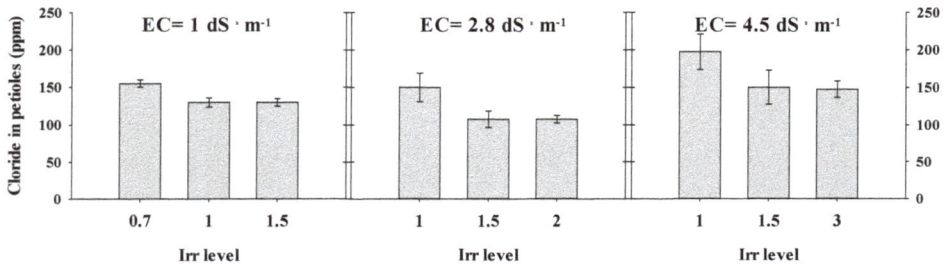

Figure 2. Chloride concentration in the petioles of pepper, treated with combinations of various salinities and irrigation (Irr) levels. Measurements were conducted in December 2015. Error bars indicate standard deviation (n = 4).

3.2. Fruit Quality

After 14 days at 7 °C and an additional three days at 20 °C, no significant differences between the treatments were observed in percentage loss of fruit weight. However, water quality, but not water quantity, affected fruit weight loss significantly (Table 2, F = 0.04). The better the water quality, the higher the weight loss (an average of 4.03%), while fruit harvested from plants irrigated at EC 2.8 dS m^{-1} had the lowest weight loss (an average of 3.55%) (Table 2). Fruit were significantly firmer (2.6 mm deformation) when irrigated with good quality water (1.6 dS m^{-1} EC), while irrigation with very salty water (EC 4.5 dS m^{-1}) gave soft fruits (3.09 mm deformation) (Table 2, F = 0.0058). The TSS was significantly affected by the water quality (Table 1, F = 0.0003); the saltier the water, the higher was the sugar content. The highest TSS content was found in the treatment of 4.5 EC − 3.0 (water salinity − amount of water. See Table 2) (8.72%), while the lowest content was found at 1.6 EC − 1.0 (water salinity − amount of water. See Table 2) (7.53%). No significant differences between the treatments were observed in percentage of decay development, although the highest decay was found in fruit irrigated with 1.6 dS m^{-1} EC (an average of 10.3% decayed fruit) and the lowest decay was found in fruit irrigated with 4.5 dS m^{-1} EC (an average of 7.1%). Water quantities did not affect all fruit quality parameters. No interaction between water quality and quantity was found in relation to external and internal fruit quality shown in Table 2.

Table 2. The influence of water quality and irrigation water amount on pepper fruit quality after 14 days at 7 °C plus three days at 20 °C. Means of six harvests during two years.

Treatment	Water Quality	Amount of Water [z]	Weight Loss (%) [y]	Flexibility (mm) [x]	TSS (%) [w]	Decay (%)
1	1.6	0.7	4.13 a [v]	2.70 a	7.58 b	14.5 a
2	1.6	1.0	4.05 a	2.58 a	7.53 b	9.3 a
3	1.6	1.5	3.90 a	2.52 a	7.55 b	7.2 a
4	2.8	1.0	3.53 a	2.32 a	7.83 ab	7.5 a
5	2.8	1.5	3.53 a	2.17 a	8.13 ab	6.0 a
6	2.8	2.0	3.58 a	2.43 a	8.12 ab	8.5 a
7	4.5	1.0	3.87 a	3.12 a	8.05 ab	7.0 a
8	4.5	2.0	3.77 a	3.02 a	8.37 ab	7.2 a
9	4.5	3.0	3.62 a	3.13 a	8.72 a	7.2 a
LSD			0.31	0.40	0.32	5.61
Mean of water quality						
	1.6		4.03 a	2.60 ab	7.56 b	10.33 a
	2.8		3.55 b	2.31 b	8.03 a	7.33 a
	4.5		3.75 ab	3.09 a	8.38 a	7.11 a
LSD			0.18	0.23	0.19	3.24
Mean of amount of water						
		Low	3.84 a	2.71 a	7.82 a	9.67 a
		Moderate	3.78 a	2.59 a	8.01 a	7.50 a
		High	3.70 a	2.69 a	8.12 a	7.61 a
LSD			0.18	0.23	0.19	3.24
Analysis of Variance (F-Value)						
WQ [u]			0.04 *	0.0058 ***	0.0003 ***	0.54 NS
AOW [t]			0.8 NS	0.86 NS	0.31 NS	0.75 NS
WA × AOW			0.97 NS	0.87 NS	0.61 NS	0.83 NS

[z] From evapo-transpiration; [y] Percentage loss from initial weight; [x] Deformation as measured in millimeters; [w] Percentage of total soluble solids (Brix°); [v] Values within each column followed by same letter(s) are not significantly different according to least significance difference test * ($p \leq 0.05$). * $p \leq 0.05$; ** $p \leq 0.01$; *** $p \leq 0.001$; **** $p \leq 0.0001$; NS = non-significant at $p \leq 0.05$; [u] Water quality (WQ); [t] Amount of water (AOW).

Vitamin C content was not affected by water quality or quantity, although fruits harvested from plants irrigated with water at EC 2.8 dS m^{-1} had the highest average vitamin C content (130 mg/100 g FW) compared with the other two water qualities (Table 3). Water quality significantly affected AOX content in the fruit after 14 days of storage and marketing simulation. The average AOX activity in fruits harvested from plants irrigated at EC of 2.8 dS m^{-1} was 4.6 μM TE/g FW compared with 4.1 and 4.0 μM TE/g FW in fruits irrigated at EC of 1.6 or 4.5 dS m^{-1}, respectively. The highest AOX activity was found in the 2.8 dS m^{-1} EC-1.5 treatment (4.8 μM TE/g FW), while the lowest activity was found in the 4.5 dS m^{-1} EC-3.0 treatment (3.9 μM TE/g FW). An interaction was found in AOX activity between the water quality and quantity (F = 0.02) (Table 3).

Table 3. Influence of water quality and irrigation water amount on fruit nutritional contents after 14 days at 7 °C plus three days at 20 °C. Means of six harvests over two years.

Treatment	Water Quality	Amount of Water	Vitamin C (mg/100 g FW)	AOX TEAC (μM TE/g FW)
1	1.6	0.7	121 a [z]	4.1 cd
2	1.6	1.0	124 a	4.1 cd
3	1.6	1.5	123 a	4.3 bcd
4	2.8	1.0	124 a	4.4 abc
5	2.8	1.5	133 a	4.8 a
6	2.8	2.0	133 a	4.6 ab
7	4.5	1.0	126 a	4.2 bcd
8	4.5	2.0	119 a	4.0 cd
9	4.5	3.0	118 a	3.9 d
LSD			10.6	0.13
Mean of water quality				
	1.6		123 a	4.1 b
	2.8		130 a	4.6 a
	4.5		121 a	4.0 b
LSD			6.13	0.08
Mean of water amount				
		Low	123 a	4.2 a
		Middle	125 a	4.3 a
		High	125 a	4.2 a
LSD			6.13	0.08
Analysis of Variance (F-Value)				
WQ [y]			0.33 NS	<0.0001 ****
AOW [x]			0.96 NS	0.76 NS
WA × AOW			0.81 NS	0.02 *

[z] Values within each column followed by same letter(s) are not significantly different according to least significance difference test ($p \leq 0.05$). * $p \leq 0.05$; ** $p \leq 0.01$; *** $p \leq 0.001$; **** $p \leq 0.0001$; NS = non-significant at $p \leq 0.05$. [y] Water quality (WQ). [x] Amount of water (AOW).

In pepper fruit, the correlation coefficient indicated a significantly higher and positive relationship between weight loss and decay development at $p = 0.01$. Weight loss had a significantly high and negative relationship with vitamin C at $p = 0.0001$. Likewise, a negative and significantly higher relationship was also noted between elasticity and decay incidence at $p = 0.01$ (Table 4).

Table 4. Correlation coefficients of weight loss (WL), elasticity (Firm), sugar content (TSS), decay, vitamin C (VC) and antioxidant activity (AOX) in red pepper after 14 days at 7 °C plus three days at 20 °C.

	WL	Firm	TSS	Decay	VC
Firm	−0.065				
TSS	−0.223	0.259			
Decay	0.342 **	−0.310 **	0.193		
VC	−0.471 ****	0.099	−0.014	−0.202	
AOX	−0.101	−0.195	0.076	−0.155	0.059

*, **, ***, and **** = significant at $p = 0.05, 0.01, 0.001$ and 0.0001 levels, respectively.

4. Discussion

Salinity and water scarcity present crucial problems for many crop species in Mediterranean countries where water resources are the main limiting factor. In these countries, the limited water quantities available to farmers and increasing water salinity impair plant growth and yield, which depend on water quantity and quality, and may vary according to the plant genotype [11,12]. Very little is known about the effect of water quantity on postharvest fruit quality, but the influence of water salinity on fruit yield and quality is well-documented; most vegetable crops have a salinity threshold at ≤2.5 dS/m [13]. Pepper plants are categorized as sensitive to moderately sensitive to salinity, although Baath et al. [14] concluded that selected chili pepper cultivars can be irrigated with water of salinity ≤3 dS/m, during at least one growing season.

We have found that water quality was more important than water quantity in determining total and export-quality yields: high water quality (1.6 dS m^{-1}) increased yield, whereas high salinity (4.5 dS m^{-1}) significantly decreased it; in both cases water quantity did not affect pepper yields. The decrease in total yield caused by salinity was mainly due to decreases in fruit fresh weight and not to the number of fruits per plant (data not shown). A high export quality fraction from the total yield was found when the salinity increased from 1 to 2.8 dS·m^{-1}. This can be explained by the lower chloride levels measured in EC 2.8 dS·m^{-1}. Previous studies found the same trend. Rameshwaran et al. [15] reported that high salinity reduced pepper yield in two growing seasons and Yasour et al. [16] reported that high water salinity reduced pepper plant biomass and fruit yield in the Arava Valley in Israel. The reduction in total yield at high salinity can be attributed to low water content in the fruit because of poor water uptake at high salt concentration, which affects cell expansion in the growing fruit [17]. It is also possible that the decreased yield and poor fruit quality associated with high salinity are caused by poor photosynthesis, which decreases CO_2 availability as a result of diffusion limitations [18], and by decreased CO_2 conductance in the stomata and mesophyll [19]. Paranychianakis and Chartzoulakis [20] reported that salt accumulation in the root zone caused development of osmotic stress and disrupted cell ion homeostasis, thereby affecting total yield. However, Urrea-Lopez et al. [21] did not find that habanero pepper fruit yield parameters were significantly affected by low photosynthetic activity associated with water salinity, probably because of the fertilizers used in their experiment.

The best fruit quality, as judged by external and internal quality parameters, after prolonged storage and shelf-life simulation, was found at a water of salinity 2.8 dS m^{-1}. Navarro et al. [22] reported that moderately saline water was beneficial when peppers were harvested at the red stage; however, no significant differences in several quality parameters were observed between irrigation with water of 1.6 dS m^{-1} and of 2.8 dS m^{-1}. It might be that plants irrigated with fresh water (1.6 dS m^{-1}) had large canopies, which evaporated more water, thereby increasing canopy humidity, which would increase postharvest decay development because of Botrytis infection (Table 1). At very high salinity (4.5 dS m^{-1}), fruit were softer and more flexible, probably because of severe disturbances in membrane permeability, water channel activity, and stomatal conductance [23]. Salinity increased sugar levels in several crops such as melons, grapes and oranges [24–26]; we have found that the saltier the water, the higher the fruit TSS. The increase in the concentration of these sugars could be due, in part, to a loss of water from the fruit, and/or in part to increased hydrolysis of sucrose, which would yield fructose and glucose, in response to the high osmotic potential in the nutrient solution. The increase in glucose and fructose concentrations could also be associated with an active osmotic adjustment [27]. The increase in sugar level in fruits harvested from high-salinity treatments also could be attributed to the increase in starch biosynthesis in developing fruits, which is believed to increase sink strength [28].

Phytonutrients such as vitamin C or AOX capacity are increasingly important aspects of fruit quality because they are associated with benefits to consumer's health [29], and pepper is considered as one of the healthier fruits [30]. Antioxidant synthesis and accumulation in plants is generally stimulated by biotic or abiotic stress such as salinity; they can protect plant organs from serious oxidative damage to lipids, proteins, and nucleic acids [22]. In the present study, the vitamin C content was not affected by water quality or quantity, but the highest vitamin C concentration was measured in water of EC 2.8 dS m^{-1}. However, in fruits harvested from plants irrigated with water of EC 2.8 dS m^{-1}, AOX was significantly higher. These results may indicate that moderate salt treatment may significantly improve the nutritional benefits of the fruit, with respect to prevention of free-radical-related diseases, as reported by Navarro et al. [22]. On the other hand, Ehret et al. [31] reported that AOX in tomato fruits responded more strongly to light and temperature than to water salinity.

In conclusion, pepper yield was increased by fresh water of good quality (EC 1.6 dS m^{-1}) and not by water quantity, whereas fruit quality after prolonged storage was better maintained in fruits irrigated with moderately saline water, of EC 2.8 dS m^{-1}. Therefore, if the water salinity does not exceed 2.8 dS/m, postharvest quality will not be impaired, although the yield will be reduced at this salinity level. However, if water quality continues to deteriorate and becomes saltier, both pepper yield and postharvest quality will be significantly affected.

Author Contributions: E.F. was the head of the project; he planned the research, analyzed the results and wrote the manuscript with the rest of the team. D.C., S.A.-T. and M.Z.-P. are research engineers in Elazar Fallik's laboratory; they conducted the experiments, evaluated fruit quality, and analyzed the data in both 2016 and 2017. R.O., S.C., and E.T. were in charge of the planning, and growing practices, yield evaluation and harvest.

Funding: This research was funded by the Chief Scientist of Israel's Ministry of Agriculture and Rural Development; project No. 430-0511-14/15/16.

Conflicts of Interest: The authors declare no conflict of interest.

References

1. Pimentel, D.; Berger, B.; Filiberto, D.; Newton, M.; Wolfe, B.; Karabinakis, E.; Clark, S.; Poon, E.; Abbett, E.; Nandaopal, S. Water resources: Agricultural and environmental issues. *BioScience* **2005**, *54*, 909–918. [CrossRef]

2. Bartels, D.; Sunkar, R. Drought and salt tolerance in plants. *Crit. Rev. Plant Sci.* **2005**, *24*, 23–58. [CrossRef]

3. Singh, S.; Grover, K.; Begna, S.; Angadi, S.; Shukla, M.; Steiner, R.; Auld, D. Physiological response of diverse origin spring safflower genotypes to salinity. *J. Arid Land Stud.* **2014**, *24*, 169–174.

4. Azuma, R.; Ito, N.; Nakayama, N.; Suwa, R.; Nguyen, N.T.; Larrinaga-Mayoral, J.A.; Esaka, M.; Fujiyama, H.; Saneoka, H. Fruits are more sensitive to salinity than leaves and stems in pepper plants (*Capsicum annuum* L.). *Sci. Hortic.* **2010**, *125*, 171–178. [CrossRef]

5. Ben-Gal, A.; Ityel, E.; Dudley, L.; Cohen, S.; Yermiyahu, U.; Presnov, E.; Zigmond, L.; Shani, U. Effect of irrigation water salinity on transpiration and on leaching requirements: A case study for bell peppers. *Agric. Water Manag.* **2008**, *95*, 587–597. [CrossRef]

6. Lee, S.K.D. Hot pepper response to interactive effects of salinity and boron. *Plant Soil Environ.* **2006**, *52*, 227–233.

7. Tripler, E.; Haquin, G.; Koch, J.; Yehuda, Z.; Shani, U. Sustainable agricultural use of natural water sources containing elevated radium activity. *Chemosphere* **2014**, *104*, 205–211. [CrossRef]

8. Suissa, A.; Silverman, D.; Friedman, O.; Tzieli, Y.; Cohen, S.; Ofenbach, R. *Recommendation to Grow Spring Pepper in the Arava*; Ministry of Agriculture and Rural Development, Extension Services, Vegetable Ward: Hanoi, Vietnam, 2017; p. 2. (In Hebrew)

9. Fallik, E.; Grinberg, S.; Alkalai, S.; Yekutieli, O.; Wiseblum, A.; Regev, R.; Beres, H.; Bar-Lev, E. A unique rapid hot water treatment to improve storage quality of sweet pepper. *Postharvest Biol. Technol.* **1999**, *15*, 25–32. [CrossRef]

10. Vinokur, Y.; Rodov, V. Method for determining total (hydrophilic and lipophilic) radical-scavenging activity in the same sample of fresh produce. *Acta Hortic.* **2006**, *709*, 53–60. [CrossRef]

11. Bie, Z.; Ito, T.; Shinohara, Y. Effects of sodium sulfate and sodium bicarbonate on the growth, gas exchange and mineral composition of lettuce. *Sci. Hortic.* **2004**, *99*, 215–224. [CrossRef]

12. Gurmani, A.R.; Khan, S.U.; Ali, A.; Rubab, T.; Schwinghamer, T.; Jilani, G.; Farid, A.; Zhang, J. Salicylic acid and kinetin mediated stimulation of salt tolerance in cucumber (*Cucumis sativus* L.) genotypes varying in salinity tolerance. *Hortic. Environ. Biotechnol.* **2018**, *59*, 461–471. [CrossRef]

13. Machado, R.M.A.; Serralheiro, R.P. Soil salinity: Effect on vegetable crop growth. Management practices to prevent and mitigate soil salinization. *Horticulturae* **2017**, *3*, 30. [CrossRef]

14. Baath, G.S.; Shukla, M.K.; Bosland, P.W.; Steiner, R.L. Irrigation water salinity influences growth stages of *Capsicum annuum*. *Agric. Water Manag.* **2017**, *179*, 246–253. [CrossRef]

15. Rameshwaran, P.; Tepe, A.; Yazar, A.; Ragab, R. The effect of saline irrigation water on the yield of pepper: Experimental and modelling study. *Irrig. Drain.* **2015**, *64*, 41–49. [CrossRef]

16. Yasour, H.; Tamir, G.; Stein, A.; Cohen, S.; Bar-Tal, A.; Ben-Gal, A.; Yermiyahu, U. Does water salinity affect pepper plant response to nitrogen fertigation? *Agric. Water Manag.* **2017**, *191*, 57–66. [CrossRef]

17. Rubio, J.S.; Garcia-Sanchez, F.; Rubio, F.; Martinez, V. Yield, blossom-end rot incidence, and fruit quality in pepper plants under moderate salinity are affected by K^+ and Ca^{2+} fertilization. *Sci. Hortic.* **2009**, *119*, 79–87. [CrossRef]

18. Flexas, J.; Diaz-Espejo, A.; Galmés, J.; Kaldenhoff, R.; Medrano, H.; Ribas-Carbo, M. Rapid variations of mesophyll conductance in response to changes in CO_2 concentration around leaves. *Plant Cell Environ.* **2007**, *30*, 1284–1298. [CrossRef]

19. Ashraf, M.; Harris, P.J.C. Photosynthesis under stressful environments: An overview. *Photosynthetica* **2013**, *51*, 163–190. [CrossRef]

20. Paranychianakis, N.V.; Chartzoulakis, K.S. Irrigation of Mediterranean crops with saline water: From physiology to management practices. *Agric. Ecosyst. Environ.* **2005**, *106*, 171–187. [CrossRef]

21. Urrea-Lopez, R.; Diaz de la Garza, R.I.; Valiente-Banuet, J.I. Effects of substrate salinity and nutrient levels on physiological response, yield, and fruit quality of habanero pepper. *HortScience* **2014**, *49*, 812–818.

22. Navarro, J.M.; Flores, P.; Garrido, C.; Martinez, V. Changes in the contents of antioxidant compounds in pepper fruits at different ripening stages, as affected by salinity. *Food Chem.* **2006**, *96*, 66–73. [CrossRef]

23. Aktas, H.; Abak, K.; Cakmak, I. Genotypic variation in the response of pepper to salinity. *Sci. Hortic.* **2006**, *110*, 260–266. [CrossRef]

24. Botia, P.; Navarro, J.M.; Cerda, A.; Martinez, V. Yield and fruit quality of two melon cultivars irrigated with saline water at different stages of development. *Eur. J. Agron.* **2005**, *23*, 243–253. [CrossRef]

25. Grieve, A.M.; Prior, L.D.; Bevington, K.B. Long-term effects of saline irrigation water on growth, yield, and fruit quality of 'Valencia' orange trees. *Aust. J. Agric. Res.* **2007**, *58*, 342–348. [CrossRef]

26. Li, X.L.; Wang, C.R.; Li, X.Y.; Yao, Y.X.; Hao, Y.J. Modifications of *Kyoho grape* berry quality under long-term NaCl treatment. *Food Chem.* **2013**, *139*, 931–937. [CrossRef] [PubMed]

27. Sato, S.; Sakaguchi, S.; Furukawa, H.; Ikeda, H. Effects of NaCl application to hydroponic nutrient solution on fruit characteristics of tomato (*Lycopersicon esculentum* Mill). *Sci. Hortic.* **2006**, *109*, 248–253. [CrossRef]

28. Petreikov, M.; Yeselson, L.; Shen, S.; Levin, I.; Schaffer, A.A.; Efrati, A.; Bar, M. Carbohydrate balance and accumulation during development of near-isogenic tomato lines differing in the AGPase-L1 allele. *J. Am. Soc. Hortic. Sci.* **2009**, *134*, 134–140.

29. Laribi, A.I.; Palou, L.; Intrigliolo, D.S.; Nortes, P.A.; Rojas-Argudo, C.; Taberner, V.; Bartual, J.; Pérez-Gago, M.B. Effect of sustained and regulated deficit irrigation on fruit quality of pomegranate cv. "Mollar de Elche" at harvest and during cold storage. *Agric. Water Manag.* **2013**, *125*, 61–70. [CrossRef]

30. Elmann, A.; Garra, A.; Alkalai-Tuvia, S.; Fallik, E. Influence of organic and mineral-based conventional fertilization practices on nutrient levels, anti-proliferative activities and quality of sweet red peppers following cold storage. *Isr. J. Plant Sci.* **2016**, *63*, 51–57. [CrossRef]

31. Ehret, D.L.; Usher, K.; Helmer, T.; Block, G.; Steinke, D.; Fret, B.; Kuang, T.; Diarra, M. Tomato fruit antioxidants in relation to salinity and greenhouse climate. *J. Agric. Food Chem.* **2013**, *61*, 1138–1145. [CrossRef]

horticulturae

MDPI

Article

The Effect of Environment and Nutrients on Hydroponic Lettuce Yield, Quality, and Phytonutrients

William L. Sublett [1], T. Casey Barickman [1,*] and Carl E. Sams [2]

[1]   North Mississippi Research and Extension Center, Department of Plant and Soil Sciences, Mississippi State University, Verona, MS 38879, USA; willsublett@gmail.com
[2]   Department of Plant Sciences, The University of Tennessee, Knoxville, TN 37996, USA; carlsams@utk.edu
*   Correspondence: t.c.barickman@msstate.edu; Tel.: +1-662-566-2201

Received: 3 October 2018; Accepted: 20 November 2018; Published: 28 November 2018

Abstract: A study was conducted with green and red-leaf lettuce cultivars grown in a deep-water culture production system. Plants were seeded in rockwool and germinated under greenhouse conditions at 25/20 °C (day/night) for 21 days before transplanting. The experimental design was a randomized complete block with a 2 × 3 factorial arrangement of cultivar and nutrient treatments that consisted of six replications. Treatments consisted of two lettuce genotypes, (1) green (Winter Density) and (2) red (Rhazes), and three nutrient treatments containing electroconductivity (EC) levels of (1) 1.0; (2) 2.0; and (3) 4.0 mS·cm^{-1}. After 50 days, plants were harvested, processed, and analyzed to determine marketable yield, biomass, plant height, stem diameter, phenolics, and elemental nutrient concentrations. An interaction between growing season and lettuce cultivar was the predominant factor influencing yield, biomass, and quality. Nutrient solution EC treatment significantly affected biomass and water content. EC treatments significantly impacted concentrations of 3-O-glucoside and uptake of phosphorous, potassium, iron, boron, zinc, and molybdenum. Effects of growing season and cultivar on leafy lettuce yield and quality were more pronounced than the effect of nutrient solution EC treatment. Thus, greenhouse production of green and red-leaf lettuce cultivars in the south-eastern United States should be conducted in the spring and fall growing seasons with elevated nutrient solution EC of ≈4.0 mS·cm^{-1} to maximize yield and quality.

Keywords: electro-conductivity; polyphenols; phenolics; flavonoids

1. Introduction

In the United States, lettuce is a valuable vegetable crop and a staple food in the diet. Lettuce contributes a notable amount of polyphenolic compounds, vitamins A, C, and E, calcium, and iron [1]. Due to its raw consumption in relatively large quantities, it provides an important source of dietary antioxidants and possesses high radical scavenging activity, which is often credited with aiding in the prevention of many chronic illnesses such as cancer and cardiovascular disease [2,3]. Lettuce is a cool-season vegetable, which thrives in temperatures ranging from 7 to 24 °C. In the southern United States, field production typically occurs in the fall and winter months, allowing growers to take advantage of shorter days and cooler temperatures. However, the increasing consumer demand for high quality, locally sourced produce and off-season availability has fueled the expansion of greenhouse production over the past decade [4]. Due to the increased ability to precisely control the greenhouse environment and maintain year-round production, lettuce yield and quality is greater, compared to open field production per unit of space [5]. The high cost of greenhouse production leaves little room for error and must be offset by high gross returns.

Southern United States greenhouse growers have production advantages during the cool seasons, such as milder temperature, greater light intensity, and reduced energy costs. Lettuce production during late spring and summer often negatively affects yield and quality and threatens economic returns [6]. In the south-east United States, adverse temperatures and long days largely limit warm season production of lettuce. Consistent exposure to these supra-optimal conditions decreases lettuce quality. For example, lettuce subjected to 13 h of daylight and temperatures above 24 °C resulted in premature inflorescence initiation, otherwise known as bolting [7]. Crisphead lettuce subjected to heat stress for a 3 or 5 day period, two weeks after heading resulted in 46% of mature lettuce heads with rib discoloration [8]. Additionally, genotype determines the susceptibility of lettuce to tipburn, but the incidence is heavily influenced by environment. An analysis of 125 harvests of butterhead lettuce over a 3-year period found that high light intensity, fresh head mass, and elevated temperature were the predominant variables positively correlated with tipburn incidence [9].

In closed greenhouse hydroponic cultivation systems, fertilizers are dissolved in water, and the total amount of solutes in the solution are referred to as the electrical conductivity (EC). Numerous studies have examined the effect of differing EC levels on lettuce production. Previous research has indicated that increasing EC levels resulted in a reduction of lettuce yield and leaf nitrate in a floating system but increased total phenolic compounds and antioxidant activity [10]. Additionally, Scuderi et al. [11] found that increasing solution EC decreased lettuce yield and resulted in reduced leaf nitrate content. Conversely, three lettuce varieties subjected to increasing EC treatments also resulted in reduced total yield but showed no significant effect on leaf nitrate content. Moreover, increasing EC levels resulted in notable increases in leaf phosphorous (P), zinc (Zn), manganese (Mn), and iron (Fe) concentrations in greenhouse lettuce [12]. While lettuce is considered mildly sensitive to high EC levels, research indicates that moderate EC is associated with the biosynthesis of secondary metabolites, such as phenolic compounds [13]. Furthermore, red-leafed lettuce varieties are characterized by higher phenolic content than green-leafed varieties. Kim et al. [14] reported that phenolic content and antioxidants increased in romaine lettuce produced with long-term irrigation and relatively low EC concentration. However, green and red-leafed baby lettuce grown with increasing EC levels contained greater amounts of flavonoids, phenolic acids, and carotenoids in both varieties [15].

Information is lacking and inconclusive regarding the effects of environmental stress on greenhouse lettuce by altering the EC of the plant nutrient solution. However, Fallovo et al. [16] investigated the effect of macro and micronutrient proportions on lettuce yield and quality of 'Green Salad Bowl' during spring and summer production seasons. The results indicated that marketable yield, leaf area index, and shoot biomass were unaffected by the nutrient solution, and growing season played the most determinant role in plant yield and quality. A high amount of calcium (Ca) did result in increased quality parameters, such as chlorophyll, glucose, fructose, and leaf Ca concentrations. Moreover, green oakleaf lettuce produced during winter and summer seasons and grown in increasing EC concentrations reached maturity more quickly during summer, and yield was unaffected regardless of nutrient solution concentration [16]. More information is needed to determine the relationship between nutrient solution EC concentrations and growing season on lettuce yield and nutritional quality. Therefore, the purpose of this study was to determine the effect of increased nutrient solution EC and growing season on lettuce plant height and stem diameter, biomass accumulation, mineral nutrient uptake, yield, and polyphenolic content of green and red-leafed lettuce cultivars.

2. Materials and Methods

2.1. Plant Culture and Harvest

Three separate studies were conducted in the spring, summer, and fall of 2016 and 2017 to examine the effects of season and nutrient solution concentrations on green and red leaf lettuce growth, minerals nutrients, and secondary metabolites. Seeds of green-leaf, 'Winter Density' lettuce, and red-leaf, 'Rhazes' lettuce, (Johnny's Selected Seed, Waterville, ME, USA) were sown into rockwool

(3.81 cm × 3.81 cm; Hummert Int., Earth City, MO, USA) and germinated in greenhouse conditions (Verona, MS, USA; 34° N, 89° W) at 25/20 °C (day/night). The natural photoperiod and light intensity were not enhanced with any supplemental lighting. Daily light intensity readings of photosynthetic active radiation (PAR) were taken using the WatchDog 1000 Series plant growth micro station (Spectrum Technologies, Aurora, IL, USA), while temperature and relative humidity were monitored with a WatchDog A-Series data logger (Spectrum Technologies, Aurora, IL, USA). After 21 days (third leaf stage), three plantlets from each cultivar were transferred into a closed hydroponic system composed of 36, 11-L Rubbermaid© Roughneck plastic storage containers (Rubbermaid, Atlanta, GA, USA). Each tub was filled with 10-L of nutrient solution using a modified Hoagland formulation [17]. Elemental concentrations of modified half-strength nutrient solution consisted of (mg·L^{-1}): N (105), P (91.5), K (117.3), Ca (80.2), Mg (24.6), S (32.0), Fe (1.0), B (0.25), Mo (0.005), Cu (0.01), Mn (0.25), and Zn (0.025). The experimental design was a randomized complete block in a 2 × 3 factorial arrangement of cultivar and EC treatments that consisted of six replications, with individual tubs representing an experimental unit. Treatments consisted of two lettuce genotypes, (1) green (Winter Density) and (2) red (Rhazes), and three nutrient treatments containing EC levels of (1) 1.0 mS·cm^{-1}; (2) 2.0 mS·cm^{-1}; and (3) 4.0 mS·cm^{-1}. Electroconductivity readings were measured weekly with a portable pH/Conductivity meter (Accumet© AP85; Fisher Scientific, Hampton, NH, USA), and growth solutions were changed every two weeks. Water was added to the containers to maintain a 10 L level of nutrient solution to keep up with the transpiration losses by the lettuce plants. After 50 days, lettuce plants were harvested by replication and treatment. Plants were separated into roots and shoots, and the fresh weights and stem diameter were recorded. A 20–30 g subsample of leaf tissue from three lettuce plants per treatment was retained to be freeze-dried (Labconco Corp., Kansas City, MO, USA). The subsamples were taken from the first fully expanded leaf of the lettuce plants. Freeze dried leaf tissue was then ground by mortar and pedestal, placed in an ultra-low freezer (−80 °C) until further analyzed for nutritional quality. The remaining plant material and roots were dried in a forced-air oven at 80 °C then weighed again to determine plant biomass production. All subsamples for chemical analysis were taken from each cultivar and treatment ($n = 3$) from each of the six replications.

2.2. Flavonoid Analysis

Flavonoid analysis was conducted according to Neugart et al. [18] and modified for the analysis of lettuce by Becker et al. [19]. Freeze-dried lettuce leaf samples were ground using a mortar and pestle for homogenous sub-samples. A 0.04 g sub-sample was extracted in a 2 mL microcentrifuge tube by adding 1.0 mL of extraction solvent (60:37:3) consisting of methanol, de-ionized water, and formic acid. The samples were then vortexed for 1 min and centrifuged at 12,000 rpm for 15 min. After centrifugation, the samples were filtered through a 0.45 μm polytetrafluoroethylene (PTFE) syringe filter and collected in a 2-mL high-performance liquid chromatography (HPLC) vial for analysis. Separation parameters and flavonoid quantification were carried out with authentic standards using an Agilent 1260 series HPLC with a multiple wavelength detector (Agilent Technologies, Willington, DE, USA). Chromatographic separations were achieved using a 150 × 4.6 mm i.d., 2.6 μm analytical scale Kinetex F5 reverse-phase column (Phenomenex, Torrance, CA, USA), which allows for effective separation of chemically similar flavonoid compounds. The column was equipped with a Kinetex F5 12.5 × 4.6 mm i.d. guard cartridge and holder (Phenomenex), and it was maintained at 30 °C using a thermostat column compartment. All separations were achieved using mobile gradient phase of reverse osmosis (RO) water adjusted to pH 2.5 with trifluoroacetic acid and acetonitrile. Anthocyanin analysis was similar to the flavonoid determination procedure with slight modifications. Briefly, 0.04 g of red lettuce sub-samples were extracted in a 2 mL microcentrifuge tube by adding 1.0 mL of extraction solvent (50:40:10) consisting of water, methanol, and acetic acid. The samples were then vortexed for 1 min and centrifuged at 12,000 rpm for 15 min. After centrifugation, the samples were filtered through a 0.45 μm PTFE syringe filter and collected in a 2 mL HPLC vial for analysis.

2.3. Mineral Composition

Nutrient analysis was conducted according to Barickman et al. [20] with slight modifications. Briefly, a 0.5 g subsample of dried leaf tissue was combined with 10-mL of 70% HNO_3, was digested in a microwave digestion unit (Model: Ethos, Milestone Inc., Shelton, CT, USA). Leaves were collected and dried for 48 h in a forced air oven (model large; Fisher Scientific, Atlanta, GA, USA) at 65 °C. Dried samples were ground to homogeneity using liquid nitrogen, and a 0.5 g sub-sample was weighed for analysis. Nutrient analysis was conducted using an inductively coupled plasma mass spectrometer (ICP-MS; Agilent Technologies, Inc., Wilmington, DE, USA). The ICP-MS system was equipped with an octopole collision/reaction cell, Agilent 7500 ICP-MS ChemStation software, a Micromist nebulizer, a water-cooled quartz spray chamber, and a CETAC (ASX-510, CETAC Inc., Omaha, NE, USA) auto-sampler. The instrument was optimized daily in terms of sensitivity (lithium: Li, yttrium: Y, thallium: Tl), level of oxide, and doubly charged ion using a tuning solution containing 10 $\mu g \cdot L^{-1}$ of Li, Y, Tl, cerium (Ce), and cobalt (Co) in a 2% HNO_3/0.5% HCl (v/v) matrix. Tissue nutrient concentrations are expressed on a dry weight (DW) basis.

2.4. Statistical Analysis

Data were subjected to the GLIMMIXED procedure and mean separation using Tukey's Honest Significant Difference test ($P \leq 0.05$) with SAS statistical software (Version 9.4; SAS Institute, Cary, NC, USA).

3. Results

3.1. Season, Cultivar, and Treatment Effects on Plant Growth and Biomass Production

Cumulative light energy levels (Figure 1A–F) registered the highest average levels in the spring and summer in both project years. Additionally, the summer growing season produced the greatest day and nighttime average temperatures in 2016 (Figure 1A–C) and 2017 (Figure 1D–F).

Statistical analysis of the results indicated that there were no effects of year (2016 and 2017). Thus, data from 2016 and 2017 were pooled and analyzed together for each lettuce plant parameter. The growing season produced a significant effect on stem diameter (Figure 2), and the lettuce cultivar impacted stem diameter (Figure 3).

The spring season produced plants with the greatest stem diameter and was statistically different than lettuce plants produced in the summer and fall season. The stem diameters of lettuce produced in the summer and fall were 32.9% and 21.3% smaller, respectively, when compared with lettuce plants produced in the spring season. Green-leaf 'Winter Density' produced plants that averaged 13.11 mm and averaged 28.5% larger stem diameter compared to red-leaf 'Rhazes' lettuce.

There were significant interactions between growing seasons and EC treatments for lettuce leaf fresh mass (FM; Figure 4). The spring season produced the greatest leaf fresh mass and was significantly more lettuce FM was produced with high and medium (4.0 and 2.0 $mS \cdot cm^{-1}$) EC treatments. There was a 17.7% increase in leaf FM when comparing the spring season, high and medium EC treatments. Conversely, there was a significant difference between spring high EC treatment leaf FM compared to the summer and fall high EC treatments. Additionally, the summer and fall high EC treatment lettuce leaf FM decreased 35.4% and 40.0%, respectively. Overall, there were significant decreases in lettuce leaf FM as the seasons progressed and EC treatments were reduced. Also, there was a significant difference between lettuce cultivars for leaf fresh mass. The green cultivar 'Winter Density' produced more fresh mass compared to the red cultivar 'Rhazes' (Figure 5). When comparing the two lettuce cultivars, there was a 42.6% decrease in lettuce fresh mass between 'Winter Density' and 'Rhazes'.

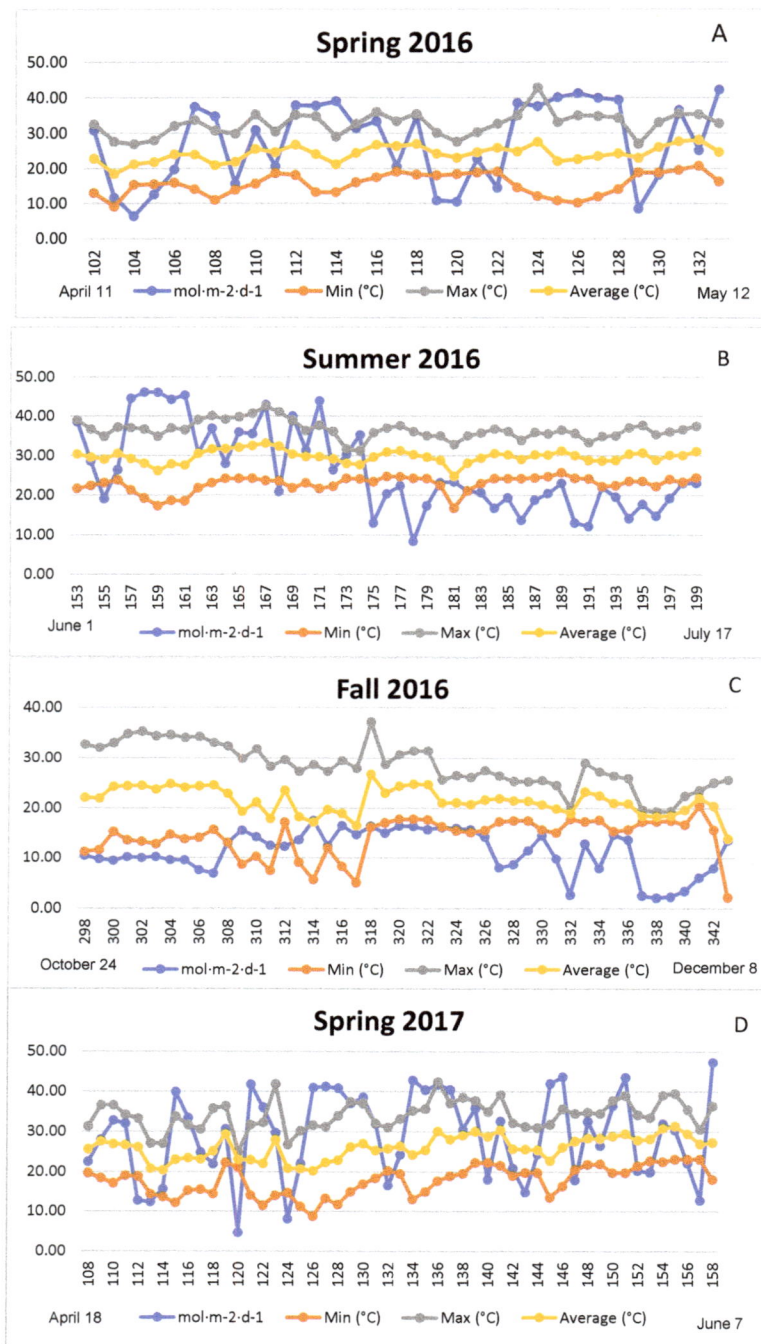

Figure 1. *Cont.*

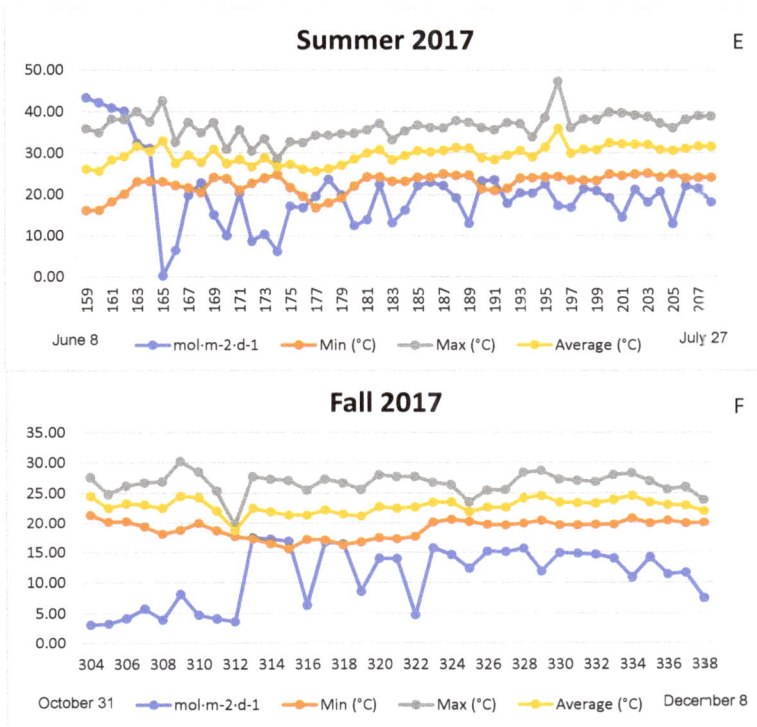

Figure 1. 2016 spring cumulative daily light energy (based on a 12 h d), and maximum, minimum, and average daily temperature (**A**); summer cumulative daily light energy, and maximum, minimum, and average daily temperature (**B**); fall cumulative daily light energy, and maximum, minimum, and average daily temperature (**C**); 2017 spring cumulative daily light energy, and maximum, minimum, and average daily temperature (**D**); summer cumulative daily light energy, and maximum, minimum, and average daily temperature (**E**); and fall cumulative daily light energy, and maximum, minimum, and average daily temperature (**F**).

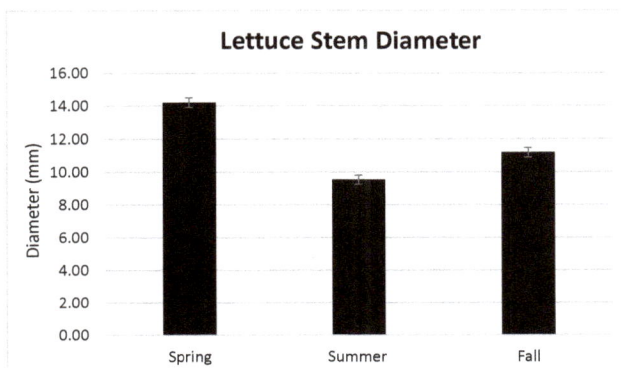

Figure 2. The effect of growing season on lettuce stem diameter. The standard error of the mean was: stem diameter ± 0.28.

Lettuce Stem Diameter

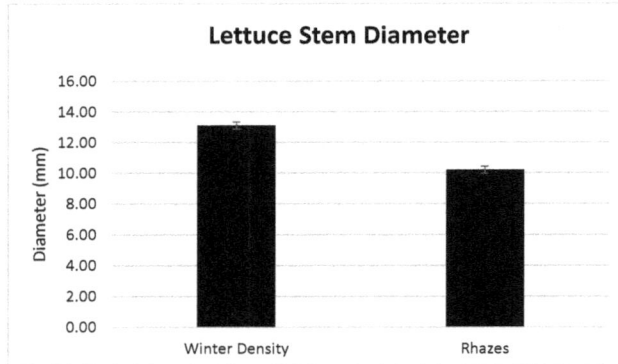

Figure 3. The effect of cultivar on greenhouse lettuce stem diameter. The standard error of the mean was: Stem diameter $\pm$ 0.23.

Leaf Fresh Mass

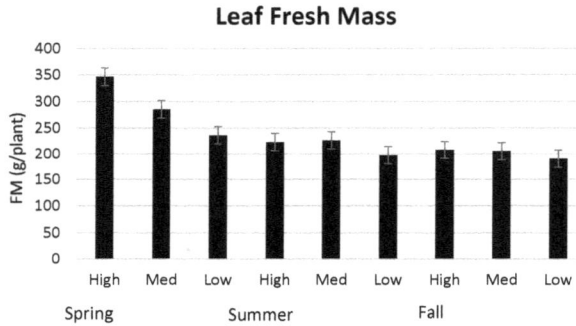

Figure 4. The interaction of growing season and electrical conductivity (EC) treatment on lettuce leaf fresh mass. The standard error of the mean: 12.65. The EC treatment: high = 4.0 mS·cm^{-1}, medium = 2.0 mS·cm^{-1}, and low = 1.0 mS·cm^{-1}.

Lettuce Fresh Mass

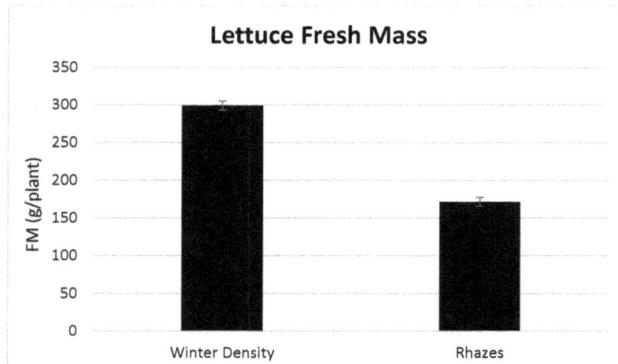

Figure 5. The effect of cultivar on greenhouse lettuce fresh mass. The standard error of the mean was: lettuce fresh mass $\pm$ 5.96.

There were no interactions between growing season, lettuce cultivar, and EC treatment for leaf dry mass (DM), DM:fresh mass (FM) ratio, and leaf water content (Table 1). Lettuce plants that were produced in the spring had significantly more leaf DM when compared to summer and fall lettuce plants. For example, there was a decrease in leaf DM by 19.4% and 33.2% when comparing the spring plants to summer and fall plants, respectively. The green-leafed lettuce cultivar 'Winter Density' produced 47.3% more leaf DM when compared to the red-leafed lettuce cultivar 'Rhazes'. Additionally, the high EC treatment produced the greatest leaf DM when compared to the medium and low EC treatments by 14.6% and 18.0%, respectively. The ratio of DM:FM was also significantly different for growing season, lettuce cultivar, and EC treatments. The summer growing season produced the greatest difference between DM:FM with a 10.2% and 10.8% increase compared to the spring and fall season, respectively. There were differences in cultivar and EC treatment DM:FM ratio. The leaf water content also saw similar trends as leaf DM in response to cultivar and EC treatment differences.

Table 1. The effect of growing season, lettuce cultivar, and EC treatment on leaf dry mass (DM) (g/plant), DM:fresh mass (FM) ratio, and lettuce leaf water content.

Treatments	Leaf DM (g)	DM:FM (g) [a]	Leaf Water %
Spring	12.37 a	0.0413 b	95.86 a
Summer	9.97 b	0.0460 a	95.41 b
Fall	8.26 c	0.0410 b	95.92 a
Winter Density	13.36 a	0.0444 a	95.56 b
Rhazes	7.04 b	0.0410 b	95.90 a
High [b]	11.44 a	0.0433 ab	95.61 b
Med	9.77 b	0.0408 b	95.93 a
Low	9.38 b	0.0440 a	95.64 b
P-Value [c,d]			
Season	***	**	***
Cultivar	***	**	**
Electro-Conductivity	**	ns	*

[a] Lettuce DM:FM is reported in grams of dry mass to grams of fresh mass; [b] The EC treatment: high = 4.0 mS·cm^{-1}, medium = 2.0 mS·cm^{-1}, and low = 1.0 mS·cm^{-1}. [c] The standard error of the mean was for growing season leaf DM ± 0.48; leaf DM:FM ± 0.0012; leaf water ± 0.13, cultivar standard error for leaf DM ± 0.40; leaf DM:FM ± 0.0011; leaf water ± 0.12, and EC treatment standard error for leaf DM ± 0.48; leaf DM:FM ± 0.0016; leaf water ± 0.13, [d] ns, *, **, *** indicate non-significant or significant at $P \leq 0.05$, 0.01, 0.001, respectively.

Lettuce root FM and DM peaked during spring production and was significantly reduced during the summer and fall concerning each cultivar (Table 2). Notably, spring green-leaf lettuce roots averaged 60.08 g FM, which was 93% greater than the root FM of red-leaf lettuce. Root biomass and water content were comparable between both cultivars produced in the fall season as well as between green-leaf lettuce grown in the summer and red-leaf lettuce grown in the spring (Table 2). Plant height (data not shown) and stem diameter were impacted and resulted in green and red-leaf summer lettuce achieving the greatest height, but smallest stem diameter, compared to their spring and summer counterparts. Rhazes lettuce growth in the fall was minimally impacted by season and cultivar and was 66% shorter compared to the Winter Density lettuce.

There were no interactions for EC treatments. Thus, EC treatments are presented as main effects. Low and high EC treatments resulted in comparable amounts of leaf DM. Conversely, lettuce leaf water content increased slightly by 0.7% when subjected to medium EC treatments. Additionally, season and treatment interactions significantly affected root biomass, water content, and stem diameter. Root biomass and water content had an inverse relationship when grown in different seasons and nutrient solution treatments. Root biomass in the spring and fall season increased by 25% and 20%, respectively, when the concentrations of the nutrient solution increased from low to high strength (data not shown). Conversely, root water content decreased 1% in the spring and fall and increased by 1% in summer with increasing nutrient solution strength. Lettuce stem diameter increased by 19% with respect to the spring season and increasing nutrient strength but decreased by 4% during the summer.

Fall production resulted in an increase of 1% from low to medium solution strength and decreased by 6% from medium to high strength.

Table 2. The effect of the interaction of season and greenhouse lettuce cultivars on root fresh mass, dry mass, dry mass to fresh mass ratio, and water content.

Season	Cultivar	Root FM (g)	Root DM (g)	Root DM:FM (g) [a]	Root Water %
Spring	Winter Density	60.08 a	2.42 a	0.04 c	0.96 a
Summer	Winter Density	30.00 b	1.47 b	0.05 b	0.95 b
Fall	Winter Density	23.73 c	1.22 c	0.05 b	0.95 b
Spring	Rhazes	31.11 b	1.32 bc	0.05 b	0.95 b
Summer	Rhazes	12.28 d	0.79 d	0.06 a	0.94 c
Fall	Rhazes	13.88 d	0.68 d	0.05 b	0.95 b
	P-Value [b,c]	***	**	**	**

[a] Lettuce DM:FM is reported in grams of dry mass to grams of fresh mass. [b] The standard error of the mean was Root FM ± 2.14; Root DM ± 0.09; Root DM:FM ± 0.002; Root Water ± 0.002; [c] ns, **, *** indicate non-significant or significant at $P \leq 0.01, 0.001$, respectively.

3.2. Season, Cultivar, and Treatment Effect on Lettuce Quality

Growing season alone demonstrated a significant effect on chlorogenic acid content of greenhouse lettuce cultivars (Figure 6). Concentrations of chlorogenic acid were statistically comparable in the spring and summer seasons but significantly different from the fall. Chlorogenic acid levels were greatest in the spring, which was 73% higher compared to the fall.

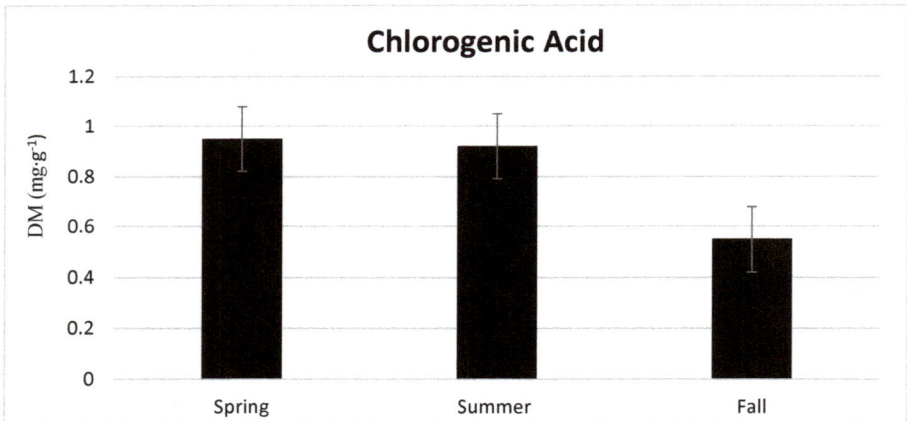

Figure 6. The effect of growing season on greenhouse lettuce chlorogenic acid content. The standard error of the mean was: Leaf DM ± 0.05. Different letters are significantly different at $P \leq 0.05$ according to Tukey's honest significant difference test.

Interactions between growing season and lettuce cultivars significantly affected chicoric acid and lettuce flavonoids (Table 3). Levels of chicoric acid increased from spring to summer to fall in both lettuce cultivars. The maximum concentration of chicoric acid, produced by red-leaf lettuce in the fall, was 131% greater compared to summer red-leaf lettuce and 175% greater than spring red-leaf lettuce.

Moreover, fall red-leaf lettuce contained 94% greater levels of chicoric acid compared to fall green-leaf lettuce. Concerning lettuce flavonoids, quercetin glucoside and quercetin glucuronide had an inverse relationship. Levels of quercetin glucoside increased from spring to summer but decreased from summer to fall in both cultivars. However, levels of quercetin glucuronide decreased from spring to summer before increasing in the fall. Spring red-leaf lettuce produced the highest concentration of luteolin (9.86 mg·g^{-1}), although maximal concentrations in green-leaf lettuce (1.56 mg·g^{-1}) were achieved in the fall. Interactions between season and cultivar resulted in increasing levels of quercetin malonyl from spring through the fall, and the greatest accumulation was present in substantially higher concentrations among red-leaf lettuce compared to green-leaf. The impact of nutrient solution treatment on lettuce phenolics was insignificant for all compounds except for quercetin glucoside, which at low-solution treatments were 69% greater than medium-solution treatments and 62% greater than high solution treatments (data not shown).

Table 3. The effect of seasons and cultivars on concentrations of greenhouse lettuce phenolics and flavonoids.

Season	Cultivar	Concentrations of Phenolics and Flavonoids (mg·g^{-1} DM) [a,b]					
		chlo	chic	qgluc	qglucor	luteolin	qmal
Spring	Winter Density	0.83 b	11.34 d	0.87 b	1.01 c	1.15 d	3.56 d
Summer	Winter Density	0.95 ab	15.26 d	1.51 b	0.55 c	0.91 d	4.63 d
Fall	Winter Density	0.52 c	33.85 b	1.05 b	1.20 c	1.56 d	5.93 cd
Spring	Rhazes	1.08 a	23.79 c	3.85 a	9.25 a	9.86 a	15.33 bc
Summer	Rhazes	0.90 ab	28.31 bc	3.98 a	5.19 b	5.40 c	18.93 b
Fall	Rhazes	0.58 c	65.52 a	1.34 b	6.50 b	7.73 b	46.90 a
	P-Value [c]	ns	**	*	*	*	***

[a] Abbreviations: chlo—chlorogenic acid; chic—chicoric acid; qgluc—quercetin glucoside; qglucor—quercetin glucuronide; qmal—quercetin malonyl; [b] The standard error of the mean was chlo $\pm$ 0.07; chic $\pm$ 3.06; qgluc $\pm$ 0.53; qglucor $\pm$ 0.64; lutein $\pm$ 0.67; qmal $\pm$ 3.57; [c] ns, *, **, *** indicate non-significant or significant at $P \le 0.05, 0.01, 0.001$, respectively.

3.3. Season, Cultivar, and Treatment Effects on Leaf Mineral Content

Growing season exhibited an effect on leaf sulfur, copper, and zinc concentrations. While the largest amount of sulfur (S) was achieved in the summer, spring growing season resulted in comparable concentrations (data not shown). Spring and fall growing seasons resulted in similar concentrations of copper (Cu), which were, respectively, 68% and 37% larger compared to the summer. Fall production resulted in the highest concentrations of zinc (Zn), followed by summer, with the lowest concentrations in the spring. Zn levels in the fall growing season were 27% greater compared to the spring. Additionally, cultivar produced a significant effect on Cu and Zn micronutrients. Both nutrients were found in the highest concentrations in the red-leaf lettuce cultivar. Cu was 33% more concentrated in red-leaf lettuce, and Zn levels were 18% larger. The interaction between season and cultivar significantly impacted the macronutrients magnesium (Mg), phosphorous (P), potassium (K), and calcium (Ca) (Table 4). Concerning green-leaf lettuce, spring production resulted in the most accumulation of Mg and Ca, which declined by 16% and 17% in the summer and an additional 7% and 12% in the fall, respectively. P and K did not display any significant changes in relation to season. Concerning red-leaf lettuce, Mg and Ca concentrations increased from spring to summer by 12% and 2%, respectively, then declined in the fall by 3% for each cultivar. P and K had the lowest accumulation in the spring (5.66/48.13 mg·g^{-1}) and steadily increased during the summer by 18% and 11% and fall season by 31% and 8%, respectively. The interaction between season and cultivar significantly impacted the micronutrients boron (B), manganese (Mn), and molybdenum (Mo) (Table 4).

Table 4. The effect of season and cultivar on the concentrations of elemental nutrients in freeze-dried greenhouse lettuce leaf tissue.

Season	Cultivar	Elemental Nutrient Concentrations [a]										
		(mg·g⁻¹) (µg·g⁻¹)										
		Mg	P	S	K	Ca	Fe	B	Mn	Cu	Zn	Mo
Spring	Winter Density	5.30 a	6.22 bc	5.22 a	48.73 c	18.02 a	120.53 abc	37.16 a	58.31 ab	4.56 a	22.39 b	0.97 a
Summer	Winter Density	4.44 b	6.46 b	5.83 a	46.26 c	14.89 b	138.57 a	28.77 b	40.59 b	2.85 b	23.58 b	0.78 b
Fall	Winter Density	4.11 bc	6.49 b	1.36 b	48.87 c	13.08 c	108.08 bc	30.50 b	47.20 b	2.83 b	24.81 b	0.51 c
Spring	Rhazes	3.34 e	5.66 c	4.39 a	48.13 c	13.83 bc	98.68 c	34.43 a	40.55 b	5.37 a	22.90 b	0.64 bc
Summer	Rhazes	3.73 cd	6.68 b	5.86 a	53.56 b	14.12 ab	128.74 ab	30.02 b	55.39 ab	3.04 b	28.36 ab	0.54 c
Fall	Rhazes	3.62 de	8.73 a	1.32 b	57.60 a	13.73 bc	127.34 ab	37.22 a	74.98 a	5.25 a	32.53 a	0.51 c
	P-Value [b,c]	***	***	ns	**	***	ns	**	*	ns	ns	*

[a] Abbreviations: Mg—Magnesium; P—Phosphorous; S—Sulfur; K—Potassium; Ca—Calcium; Fe—Iron; B—Boron; Mn—Manganese; Cu—Copper; Zn—Zinc; Mo—Molybdenum; [b] The standard error of the mean was Mg ± 0.15; P ± 0.27; S ± 0.70; K ± 1.63; Ca ± 0.55; Fe ± 14.53; B ± 1.43; Mn ± 9.35; Cu ± 0.66; Zn ± 2.64; Mo ± 0.68; [c] ns,*,**,*** indicate non-significant or significant at $P < 0.05$, 0.01, 0.001, respectively.

Concerning green-leaf lettuce, B and Mn concentrations were greatest in the spring; whereas, summer and fall concentrations did not significantly differ. Mo concentrations were greatest in the spring (0.97 µg·g^{-1}) and decreased during the summer by 20% and an additional 35% in the fall. Concerning red-leaf lettuce, B and Mn concentrations were greatest in the fall. However, B concentration decreased 13% from spring to summer, while Mn increased 37% from spring to summer. Molybdenum concentrations decreased 16% from spring to summer and an additional 6% from summer to fall. Increasing solution EC impacted leaf concentrations of P, K, Fe, B, Zn, and Mo. Each nutrient increased from treatment 1 to treatment three except for K, which reached a saturation point at treatment 2 and declined with the elevated EC of treatment 3. Additionally, this general trend was observed concerning the other mineral nutrients that were considered not statistically significant.

4. Discussion

4.1. Season, Cultivar, and Treatment Effect on Plant Growth and Biomass Production

The current study examines how the seasonal environment and increasing nutrient solution EC affect lettuce root and shoot mass, plant height and stem diameter, mineral nutrient content, and concentrations of selected phenolic compounds in green and red-leaf romaine cultivars. While season, cultivar, and EC treatments created significant differences in leaf fresh mass and stem diameter, it was the interaction between growing season and lettuce cultivar that demonstrated the most significant effect on root and shoot biomass. Spring growing season and highest EC treatment resulted in the greatest production of leaf and root FM in both cultivars. Greenhouse environmental data measured during 2016 and 2017 show that the spring growing seasons registered the highest levels of cumulative light energy. Light is known as a primary regulatory factor in plant growth and development, and previous research has indicated that daily light intensity significantly affects the production of shoot biomass. For example, Fu et al. [21] examined the effect of increasing light intensity (60, 140, and 220 µmols·m^{-2}·s^{-1}) and nitrogen concentrations (7, 15, and 23 mmols·L^{-1}) on the growth and quality of hydroponic leaf lettuce. The results revealed that plants subjected to 220 µmols·m^{-2}·s^{-1} light intensity and 7 mmols·L^{-1} of N produced the greatest amount of dry biomass. Similarly, lettuce plants grown during fall and spring seasons with 50 or 100 µmols·m^{-2}·s^{-1} of supplemental white light produced more than 270% greater biomass production compared to control treatments [22].

The current study's results suggest that lettuce cultivar had the greatest influence on the production of leaf FM content in green and red-leafed cultivars. Lettuce leaf DM, DM:FM ratio, and leaf water content were influenced the most by growing season, lettuce cultivar, and EC treatments. There was an interaction between the growing season and lettuce cultivars that created the most consistent favorable conditions for the production of root biomass, root DM:FM ratio, and root water content. These results are mixed with other studies that demonstrate lettuce sensitivity to increasing EC concentrations [12]. In the spring, EC treatments were significantly greater compared to the summer and fall growing seasons. Consequently, the summer and fall growing season correspond to other studies. For example, Scuderi et al. [11] reported that increasing solution EC decreased yield and leaf nitrate content in lettuce planted at high densities in a deep-water culture production system. Furthermore, previous research demonstrated that increasing salinity treatments in three lettuce cultivars also resulted in reduced total yield [12]. Temperature is known to heavily influence the partitioning of photoassimilates in plants, and studies of lettuce [16,23], tomato [4], and zucchini [24] have indicated differences in plant biomass due to light and temperature interactions. Under suboptimal conditions, lettuce's resilience to common physiologically induced disorders such as tipburn [9,23,25], rib-discoloration [26], bolting [7], and the increase of bitterness compounds [27] is highly correlated to lettuce genotype.

4.2. Season, Cultivar, and Treatment Effect on Lettuce Quality

Previous research has demonstrated that despite the influence on lettuce yield, increasing EC levels caused greater production of flavonoid and phenolic compounds [14,15]. The results of the current study were inconsistent with these findings. Nutrient solution EC did not significantly affect flavonoid and phenolic concentration of any compounds except for quercetin glucoside, which was the highest flavonoid concentration in the leaf tissue and grown under the lowest EC treatment. However, season and the interaction between season and lettuce cultivar showed a significant impact on phenolic production. Chlorogenic acid is well studied in plants and acts as an antioxidant as well as protecting against ultra-violet radiation [28]. This corresponds with the results of the current study, indicating the greatest concentrations of chlorogenic acid in the spring and summer when greenhouse light intensity was at its peak. Furthermore, red-leaf lettuce cultivars contain higher concentrations of phenolic compounds than their green-leaf counterparts, and previous studies have shown great variability in the production of these compounds with respect to cultivar and growth environment. For example, Oh et al. [29] reported that exposing five-week-old lettuce plants to mild environmental stresses resulted in a two to three-fold increase in phenolic compounds in the leaf tissue. Specifically, the study found that decreasing temperature elevated concentrations of quercetin and luteolin glycosides. Moreover, increasing photosynthetic photon flux density (PPFD) from 43 to 410 μmols·m^{-2}·s^{-1} also increased concentrations of quercetin, luteolin, and cyanidin glycosides [19], and increasing ultraviolet (UV) radiation in field grown lettuce resulted in a dose-dependent response of quercetin and luteolin glycosides and total phenolic acid concentrations [28]. These findings are consistent with the results of the current study, which demonstrated significant increases in flavonoids and phenolic content among red and green-leaf cultivars during spring and fall growing seasons where PPFD levels were higher and average daily temperatures were cooler, respectively, compared to summer.

4.3. Season, Cultivar, and Treatment Effects on Leaf Mineral Content

While climatic factors predominantly influenced the content of lettuce flavonoid and phenolic compounds, all production variables in the current study affected the uptake and concentration of leaf mineral nutrients. In field production, the uptake of mineral nutrients occurs when nutrients become available, which is dependent on soil pH, buffering capacity, and moisture [30]. It is generally accepted that increasing the nutrient supply when nutrients are already present in sufficient amounts will not improve plant growth, especially under extreme adverse environmental conditions [31]. However, in hydroponic production systems, plant roots are provided with a constant supply of purified water with a low buffering capacity. The pH of this water can be adjusted and held at the preferred range of 5.5 to 6.0, which allows maximum availability of nutrients to plant roots. Previous research indicated that even slight increases of pH to levels of 7.0 could significantly reduce lettuce FM and DM [32]. Several studies have examined the effect of increased nutrient solution EC on plant mineral nutrient content. Fallovo et al. [16] investigated the effect of growing season and increasing nutrient solution EC on yield and quality of hydroponic lettuce. The results of this study demonstrated that leaf mineral content of macroelements P, K, and Mg increased with increasing solution EC. Additionally, altering macro-anion and macro-cation nutrient solution proportions in spring and summer growing seasons significantly affected leaf concentrations of N, K, Mg, and Ca [16]. Furthermore, Barickman et al. [30] found that elevating K for greenhouse lettuce production resulted in higher concentrations of K in lettuce leaf tissue. However, a saturation point was reached before negative effects developed at higher levels of K fertilization. The results of these experiments are consistent with the findings of the current study where season, cultivar, and the interactions between the two demonstrated the most significant effect on leaf mineral nutrient content. Additionally, mineral nutrient concentrations increased with increasing solution EC except for K, which reached a saturation point and decreased in plants exposed to the highest solution concentration.

Horticulturae **2018**, *4*, 48

To develop a thorough understanding of the genotypical mechanisms and external contributing factors that produce variable results with respect to lettuce growth and development, secondary compound production, and sequestration of mineral nutrients, more information is required. While it is generally true that exposing lettuce to mild abiotic stresses, specifically elevated light irradiance and temperature, the effects of increasing growth solution EC are inconsistent concerning yield and quality. While the results of this study agree with previous work that suggested yield and quality are predominantly affected by growing season as opposed to increasing EC, all the tested leaf elemental nutrient concentrations increased as nutrient solution EC increased with statistical significance. Thus, the results of this study suggest that fall and spring production of greenhouse green and red-leaf cultivars with elevated EC solution of ≈ 4.0 mS·cm^{-1} should be used to maximize lettuce yield and nutritional quality.

Author Contributions: For this research article, T.C.B. conceived and designed the experiments; W.L.S. performed the experiments; T.C.B. and W.L.S. analyzed the data; T.C.B., W.L.S., and C.E.S. contributed sample analysis; W.L.S. wrote the manuscript; T.C.B. and C.E.S. edited the manuscript.

Funding: This research received no external funding.

Acknowledgments: This publication is a contribution of the Mississippi Agriculture and Forestry Experiment Station and supported by the USDA NIFA Hatch S-294 Project MIS 146030. This research received no external funding.

Conflicts of Interest: The authors declare no conflict of interest.

References

1. Romani, A.; Pinelli, P.; Galardi, C.; Sani, G.; Cimato, A.; Heimler, D. Polyphenols in Greenhouse and Open-Air-Grown Lettuce. *Food Chem.* **2002**, *79*, 337–342. [CrossRef]
2. Husain, S.R.; Cilurd, J.; Cillard, P. Hydroxyl radical scavenging activity of Flavonoids. *Phytochemistry* **1987**, *26*, 2489–2491. [CrossRef]
3. Cartea, M.E.; Francisco, M.; Soengas, P.; Velasco, P. Phenolic Compounds in Brassica Vegetables. *Molecules* **2011**, *16*, 251–280. [CrossRef] [PubMed]
4. Gruda, N. Impact of environmental factors on product quality of greenhouse vegetables for fresh consumption. *Crit. Rev. Plant Sci.* **2005**, *24*, 227–247. [CrossRef]
5. Jensen, M. Hydroponics Worldwide. *Acta Hortic.* **1999**, *481*, 719–729. [CrossRef]
6. Zbeetnoff, C. *The North American Greenhouse Vegetable Industry. Farm Credit Canada and AgriSuccess*; Agro-Environmental Consulting: White Rock, BC, Canada, 2006.
7. Rappaport, L.; Wittwer, S.H. Night temperature and photoperiod effects on flowering of leaf lettuce. *Proc. Am. Soc. Hortic. Sci.* **1956**, *68*, 279–282.
8. Jenni, S. Rib Discoloration: A Physiological Disorder Induced by Heat Stress in Crisphead Lettuce. *HortScience* **2005**, *40*, 2031–2035.
9. Wissemeier, A.H.; Zühlke, G. Relation between climatic variables, growth and the incidence of tipburn in field-grown lettuce as evaluated by simple, partial and multiple regression analysis. *Sci. Hortic.* **2002**, *93*, 193–204. [CrossRef]
10. Zanin, G.; Ponchia, G.; Sambo, P. Yield and quality of vegetables grown in a floating system for readyto-eat produce. *Acta Hortic.* **2009**, *807*, 433–438. [CrossRef]
11. Scuderi, D.; Restuccia, C.; Chisari, M.; Barbagallo, R.N.; Caggia, C.; Giuffrida, F. Salinity of nutrient solution influences the shelf-life of fresh-cut lettuce grown in floating system. *Postharvest Biol. Technol.* **2011**, *59*, 132–137. [CrossRef]
12. Abou-Hadid, A.F.; Abd-Elmoniem, E.M.; El-Shinawy, M.Z.; Abou-Elsoud, M. Electrical conductivity effect on growth and mineral composition of lettuce plants in hydroponic system. *Acta Hortic.* **1996**, *434*, 59–66. [CrossRef]
13. Sgherri, C.; Perez-Lopez, U.; Micaelli, F.; Miranda-Apodaca, J.; Mena-Petite, A.; Munoz-Rueda, A.; Quartacci, M.F. Elevated CO_2 and salinity are responsible for phenolics-enrichment in two differently pigmented lettuces. *Plant Physiol. Biochem.* **2017**, *115*, 269–278. [CrossRef] [PubMed]

14. Kim, H.J.; Fonseca, J.M.; Choi, J.H.; Kubota, C.; Dae, Y.K. Salt in irrigation water affects the nutritional and visual properties of romaine lettuce (*Lactuca sativa* L.). *J. Agric. Food Chem.* **2008**, *56*, 3772–3776. [CrossRef] [PubMed]
15. Neocleous, D.; Koukounaras, A.; Siomos, A.S.; Vasilakakis, M. Assessing the salinity effects on mineral composition and nutritional quality of green and red 'baby' lettuce. *J. Food Qual.* **2014**, *37*, 1–8. [CrossRef]
16. Fallovo, C.; Rouphael, Y.; Cardarelli, M.; Rea, E.; Battistelli, A.; Colla, G. Yield and quality of leafy lettuce in response to nutrient solution composition and growing season. *J. Food Agric. Environ.* **2009**, *7*, 456–462.
17. Hoagland, D.R.; Arnon, D.I. The water-culture method for growing plants without soil. *Calif. Agric. Exp. Stn. Circ.* **1950**, *347*, 1–32.
18. Neugart, S.; Zietz, M.; Schreiner, M.; Rohn, S.; Kroh, L.W.; Krumbein, A. Structurally different flavonol glycosides and hydroxycinnamic acid derivatives respond differently to moderate UV-B radiation exposure. *Physiol. Plant.* **2012**, *145*, 582–593. [CrossRef] [PubMed]
19. Becker, C.; Kläring, H.P.; Kroh, L.W.; Krumbein, A. Temporary reduction of radiation does not permanently reduce flavonoid glycosides and phenolic acids in red lettuce. *Plant Physiol. Biochem.* **2013**, *72*, 154–160. [CrossRef] [PubMed]
20. Barickman, T.C.; Kopsell, D.A.; Sams, C.E. Selenium influences glucosinolate and isothiocyanates and increases sulfur uptake in Arabidopsis thaliana and rapid-cycling Brassica oleracea. *J. Agric. Food. Chem.* **2013**, *61*, 202–209. [CrossRef] [PubMed]
21. Fu, Y.; Li, H.; Yu, J.; Liu, H.; Cao, Z.; Manukovsky, N.S.; Liu, H. Interaction effects of light intensity and nitrogen concentration on growth, photosynthetic characteristics and quality of lettuce (*Lactuca sativa* L. Var. youmaicai). *Sci. Hortic.* **2017**, *214*, 51–57. [CrossRef]
22. Gaudreau, L.; Charbonneau, J.; Canda, A.; Gv, Q. Photoperiod and Photosynthetic Photon Flux Influence Growth and Quality of Greenhouse-grown Lettuce. *HortScience* **1994**, *29*, 1285–1289.
23. Glenn, E.P. Seasonal effects of radiation and temperature on growth of greenhouse lettuce in a high insolation desert environment. *Sci. Hortic.* **1984**, *22*, 9–21. [CrossRef]
24. Rouphael, Y.; Colla, G. Growth, yield, fruit quality and nutrient uptake of hydroponically cultivated zucchini squash as affected by irrigation systems and growing seasons. *Sci. Hortic.* **2005**, *105*, 177–195. [CrossRef]
25. Bres, W.; Weston, L. A Nutrient Accumulation and Tipburn in NFT-grown Lettuce at Several Potassium and pH Levels. *HortScience* **1992**, *27*, 790–792.
26. Jenni, S.; Truco, M.J.; Michelmore, R.W. Quantitative trait loci associated with tipburn, heat stress-induced physiological disorders, and maturity traits in crisphead lettuce. *Theor. Appl. Genet.* **2013**, *126*, 3065–3079. [CrossRef] [PubMed]
27. Bunning, M.L.; Kendall, P.A.; Stone, M.B.; Stonaker, F.H.; Stushnoff, C. Effects of Seasonal Variation on Sensory Properties and Total Phenolic Content of 5 Lettuce Cultivars. *J. Food Sci.* **2010**, *75*, 156–161. [CrossRef] [PubMed]
28. García-Macías, P.; Ordidge, M.; Vysini, E.; Waroonphan, S.; Battery, N.H.; Gordon, M.H.; Hadley, P.; John, P.; Lovegrove, J.A.; Wagstaffe, A. Changes in the flavonoid and phenolic acid contents and antioxidant activity of red leaf lettuce (Lollo Rosso) due to cultivation under plastic films varying in ultraviolet transparency. *J. Agric. Food Chem.* **2007**, *55*, 10168–10172. [CrossRef] [PubMed]
29. Oh, M.; Carey, E.E.; Rajashekar, C.B. Plant Physiology and Biochemistry Environmental stresses induce health-promoting phytochemicals in lettuce. *Plant Physiol. Biochem.* **2009**, *47*, 578–583. [CrossRef] [PubMed]
30. Barickman, T.C.; Horgan, T.E.; Wheeler, J.R.; Sams, C.E. Elevated Levels of Potassium in Greenhouse-grown Red Romaine Lettuce Impacts Mineral Nutrient and Soluble Sugar Concentrations. *HortScience* **2016**, *51*, 504–509.
31. Hu, Y.; Schmidhalter, U. Drought and salinity: A comparison of their effects on mineral nutrition of plants. *J. Plant Nutr. Soil Sci.* **2005**, *168*, 541–549. [CrossRef]
32. Anderson, T.S.; Martini, M.R.; de Villiers, D.; Timmons, M.B. Growth and Tissue Elemental Composition Response of Butterhead Lettuce (*Lactuca sativa*, cv. Flandria) to Hydroponic Conditions at Different pH and Alkalinity. *Horticulturae* **2017**, *3*, 41. [CrossRef]

horticulturae

MDPI

Article

Monitoring Dormancy Transition in Almond [*Prunus Dulcis* (Miller) Webb] during Cold and Warm Mediterranean Seasons through the Analysis of a *DAM* (*Dormancy-Associated MADS-Box*) Gene

Ángela S. Prudencio, Federico Dicenta and Pedro Martínez-Gómez *

Department of Plant Breeding, CEBAS-CSIC (Centro de Edafología y Biología Aplicada del Segura-Consejo Superior de Investigaciones Científicas), PO Box 164, 30100 Espinardo, Murcia, Spain; asanchez@cebas.csic.es (A.S.P.); fdicenta@cebas.csic.es (F.D.)

* Correspondence: pmartinez@cebas.csic.es; Tel.: +34-968-396-200

Received: 8 October 2018; Accepted: 8 November 2018; Published: 19 November 2018

Abstract: For fruit tree (*Prunus*) species, flower bud dormancy completion determines the quality of bud break and the flowering time. In the present climate change and global warming context, the relationship between dormancy and flowering processes is a fundamental goal in molecular biology of these species. In almond [*P. dulcis* (Miller) Webb], flowering time is a trait of great interest in the development of new cultivars adapted to different climatic areas. Late flowering is related to a long dormancy period due to high chilling requirements of the cultivar. It is considered a quantitative and highly heritable character but a dominant gene (*Late bloom, Lb*) was also described. A major QTL (quantitative trait loci) in the linkage group (LG) 4 was associated with *Lb*, together with other three QTLs in LG1 and LG7. In addition, *DAM* (*Dormancy-Associated MADS-Box*) genes located in LG1 have been largely described as a gene family involved in bud dormancy in different *Prunus* species including peach [*P. persica* (L.) Batsch] and Japanese apricot (*P. mune* Sieb. et Zucc.). In this work, a *DAM* transcript was cloned and its expression was analysed by qPCR (quantitative Polymerase Chain Reaction) in almond flower buds during the dormancy release. For this purpose two almond cultivars ('Desmayo Largueta' and 'Penta') with different chilling requirements and flowering time were used, and the study was performed along two years. The complete coding sequence, designated *PdDAM6* (*Prunus dulcis DAM6*), was subjected to a phylogenetic analysis with homologous sequences from other *Prunus* species. Finally, expression dynamics analysed by using qPCR showed a continuous decrease in transcript levels for both cultivars and years during the period analysed. Monitoring almond flower bud dormancy through *DAM* expression should be used to improve almond production in different climate conditions.

Keywords: flowering; breeding; chilling requirements; qPCR; transcription; cloning

---

1. Introduction

During autumn temperate fruit trees (*Prunus*) activate a survival strategy called endodormancy, to protect against unfavourable chill conditions. Trees cease growth and form structures called buds in order to protect meristems from unfavourable environmental conditions, including low temperature and desiccation [1]. Chill accumulation allows the progression from flower bud endodormancy stage to flower bud ecodormancy which is regulated by heat accumulation [2]. Flowering time in almond [*Prunus dulcis* (Miller) Webb] is mainly dependent on chilling requirements to overcome this endodormancy stage [3]. These chilling requirements are considered a cultivar-dependent trait, correlated with species or cultivar origin [4,5].

Warm winter temperatures affect cold accumulation and if chilling requirement is not fully satisfied such a condition could lead to irregular and insufficient flowering with a loss of production [4,6]. Due to its economic importance, dormancy release is being studied in different species, although knowledge is still scarce and no common mechanism has been described. Thus, expression analysis of candidate genes may be a useful tool for the interannual monitoring of endodormancy progression within the flower bud. This is especially interesting for commercial fruit tree cultivars displaying a wide range of flowering and ripening time phenotypes, as in case of almond [6]. Adaptation to climatic conditions largely depends on an adequate flowering time, and it is one of the most important agronomic traits in almond breeding programs, as it determines whether the pollination period will occur in favourable climatic conditions [4]. In this context, the development of new extra-late flowering cultivars to avoid the spring frosts is one of the main objectives of almond breeding programs [7].

Late flowering is considered a quantitative and highly heritable character but a major gene (*Late bloom*, *Lb*) was also described. A QTL (quantitative trait loci) explaining 57% of the observed variance in the Linkage Group (LG) 4 was associated with *Lb*, together with other three QTLs (explaining 20, 12, and 8% of the variance) in LG1 and LG7 [8,9]. In addition, bud endodormancy has a set of genetic controls which may be characterized through examination of gene expression in bud tissues over time. Prior studies showed the *Dormancy Associated MADS-Box* (*DAM*) gene family is a group of transcription factors that regulated peach [*P. persica* (L.) Batsch] dormancy [10]. This gene family was discovered in the *Evergrowing* (*evg*) mutant of peach. The mutation consisted of a deletion in the *EVG* (*EVERGROWING*) locus affecting up to four genes which prevents terminal buds from entering the endodormancy stage [11]. The map-based cloning analyses of *EVG* locus revealed that it included six tandemly arrayed genes [11,12]. Moreover, *DAM5* expression was analysed in different cultivars of peach [13] and *PpDAM6* was postulated as one of the main factors involved in the regulation of dormancy in different *Prunus* species including peach [14–16] and Japanese apricot (*P. mume* Sieb. et Zucc.) [17,18].

In this work, a candidate *DAM* transcript was cloned and expression from endodormancy to ecodormancy stages in two almond cultivars with different chilling requirements and flowering time: 'Desmayo Largueta' and 'Penta' was determined.

2. Materials and Methods

2.1. Plant Material

'Desmayo Largueta' is a traditional Spanish almond cultivar with very low chilling requirements and extra-early flowering time, and 'Penta', a cultivar released from the Almond Breeding Program of CEBAS-CSIC (Murcia, South-East Spain) with high chilling requirements and extra-late flowering time, were used [6]. The plant material consisted of flower buds sampled weekly between stages A (dormancy phase) and B (after dormancy release) referenced to the phenological stages described by Felipe [19] (Figure 1).

2.2. Chilling Requirements Evaluation

Experiments for the evaluation of chilling and heat requirements were conducted in the experimental field of CEBAS-CSIC, in Murcia (South-East Spain), during two seasons of study: 2015–2016 and 2016–2017. Temperatures were recorded hourly with a data logger (HOBO® UX100-003 Temp/Relative Humidity, Madrid, Spain) from November to February during both seasons. Three branches (40 cm in length and 5 mm in diameter) were collected weekly from the same tree in the field, and placed in a growth chamber in controlled conditions (25 ± 1 °C, RH 40 ± 3.5% during a 16 h light photoperiod and 20 ± 1 °C, RH 60% during the dark period). Almond branches were placed in the growth chamber, in a 5% sucrose solution and 1% aluminium sulphate, making a fresh cut in the base of the branches. After 10 days, the development state of the flower buds was recorded.

The date of dormancy breakage was established when, after 10 days in the growth chamber, 50% of the flower buds were in the B-C state [20]. The calculation of chilling accumulation in field conditions was calculated as the chill contributions in the field necessary for breaking of dormancy (transition from stage A to stage B, see Figure 1) in chill units (CUs) according to the model described by Richardson et al. [16] with an initial date for chilling accumulation when consistent chilling accumulation occurred and temperatures producing a negative effect (chilling negation) were rare [21]. These CUs were calculated as hours below 7 °C. In addition, chilling accumulation was calculated in chill portions (CPs) according to the dynamic model [22,23] with an initial chilling accumulation of 0. The model is based on the assumption that dormancy completion may be estimated as a dynamic two-stage process controlling an accumulated bud break factor. The model is "dynamic" in the sense that relatively high temperatures, typically 19 °C and above, effectively negate earlier chilling; alternatively, moderate temperatures, typically around 13–14 °C effectively enhance moderate earlier chilling temperatures.

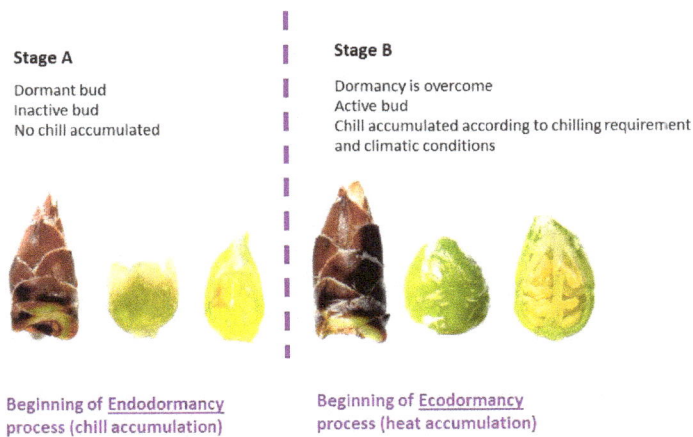

Figure 1. Plant material assayed from the almond cultivar 'Desmayo Largueta'. Flower buds in the dormant stage (A) and after dormancy release (B).

2.3. cDNA Isolation and Cloning

Almond samples assayed include flower buds in state A (completely dormant bud), state B (when 40–50% of the chilling requirement of each cultivar are satisfied) and state B (when the flower bud has broken its dormancy) (Figure 1). Total RNA was extracted from almond flower buds [24] and treated with DNAseI (Ambion). cDNA was synthetized using SSIII Reverse Transcriptase (ThermoFisher Scientific, Waltham, MA, USA). The full-length cDNA was isolated from cDNA of flower buds of 'Desmayo Largueta' and 'Penta' cultivars using 3'-RACE strategy and specific primers from *Prunus persica* available sequences in databases. High-fidelity PCR (Polymerase Chain Reaction) was performed using KOD (from Archaeon Thermococcus kodakaraensis) Hot Start DNA polymerase (Novagen, Berlin, Germany) and the product was cloned into *E. coli* using Zero Blunt Topo PCR Cloning Kit (Life Technologies, Carlsbad, CA, USA) for sequencing.

2.4. Phylogenetic Analysis

A BLAST (Basic Local Alignment Search Tool) search was performed with the full-length PdDAM6 cDNA (*Prunus Dulcis DORMANCY-ASSOCIATED MADS-BOX 6*), in order to confirm the identity of the sequence and to collect homologous proteins from the Prunus genus with a high percentage identity. A pPutative PdDAM6 protein sequence was obtained using the ExPASY translate tool (http://web.expasy.org/translate/). A phylogenetic tree was created using Philogeny.fr (http://phylogeny.lirmm.fr/phylo_cgi/index.cgi).

2.5. Gene Expression Analysis

To investigate the expression pattern of PdDAM6 during bud dormancy progression, real time qPCR experiments were executed with a One Step Plus real-time PCR system (Applied Biosystems, Foster City, CA, USA). Specific primers were designed based on an almond PdDAM6 sequence using Primer3 software (Forward Primer: 5′ AGGAAATACTGGACCTGCGT-3′; Reverse Primer: 5′-GGTGGAGGTGGCAATTATGG-3′). Reaction efficiency was checked by the standard curve method. For all real-time qPCR reactions, a 10 µL mix was made including: 5 µL Power SYBR®Green PCR Master Mix (Applied Biosystems, Foster City, CA, USA), 20 ng of cDNA, and 0.5 µL of each primer (5 µM). High-fidelity PCR was performed using KOD Hot Start DNA polymerase (Novagen, Berlin, Germany), and the product was cloned into *E. coli* using a Zero Blunt Topo PCR Cloning Kit (Life Technologies, Carlsbad, CA, USA) for sequencing. PCR was performed in a 30 µL mix according to the manufacturer's instructions with 150 ng of cDNA from each almond cultivar and 10 µM primers. The PCR reaction was incubated at 94 °C for 2′ for the initial denaturalisation step, followed by 35 cycles of 94 °C for 30″, 62 °C for 1′ and 68 °C for 1′. A final extension step at 68 °C was set for 10′. Each biological sample was implemented in duplicate. RPII was used as reference gene for data normalization using primers designed by Tong et al. [25] (Forward Primer: 5′-TGAAGCATACACCTATGATGATGAAG; Reverse primer: 5′ CTTTGACAGCACCAGTAGATTCC-3′) and the levels of relative expression were calculated by the 2−ΔΔCt method [26] taking Ct value from November the 10th samples as the reference expression level.

3. Results

3.1. Chilling Requirements Evaluation

Chilling accumulation in field conditions during the two seasons of study (2015–2016 and 2016–2017) calculated as chill units (CUs) according to the Richardson model and in chill portions (CPs) according to the dynamic model is shown in Figure 2.

During the first year of the study (2015–2016), an important reduction of chill accumulation was observed, mainly in terms of chill units (CUs). As shown in Table 1, the dormancy release date was observed earlier during the 2016–2017 season, when a higher amount of chill units accumulated. Regarding flowering time, an important advance was observed in the late cultivar 'Penta'.

Table 1. Chill accumulated percentage (chill portions) in field conditions during the seasons 2015–2016 and 2016–2017.

Season		Stage A		Stage B		Flowering Date
		Date	Chill Accumulation	Date	Chill Accumulation	
'Desmayo Largueta'	2015/2016	November 10	0	December 21	16	January 28
	2016/2017	November 10	0	December 15	24	January 27
'Penta'	2015/2016	November 10	0	February 10	41	March 25
	2016/2017	November 10	0	February 2	54	March 12

3.2. cDNA Isolation, Cloning and Phylogenetic Analysis

Phylogenetic analysis of PdDAM6 protein sequence from 'Desmayo Largueta' and 'Penta' sequences confirmed that *PdDAM6* is indeed a member of the DAM family transcription factors. In addition, the phylogenetic tree showed that PdDAM6 branch (including sequences from 'Desmayo Largueta' and 'Penta' cultivars) is closer to *Prunus persica* DAM6 (*PpDAM6*) and *Prunus pseudocerasus* (*PpsDAM6*) rather than to DAM5 protein group (Figure 3).

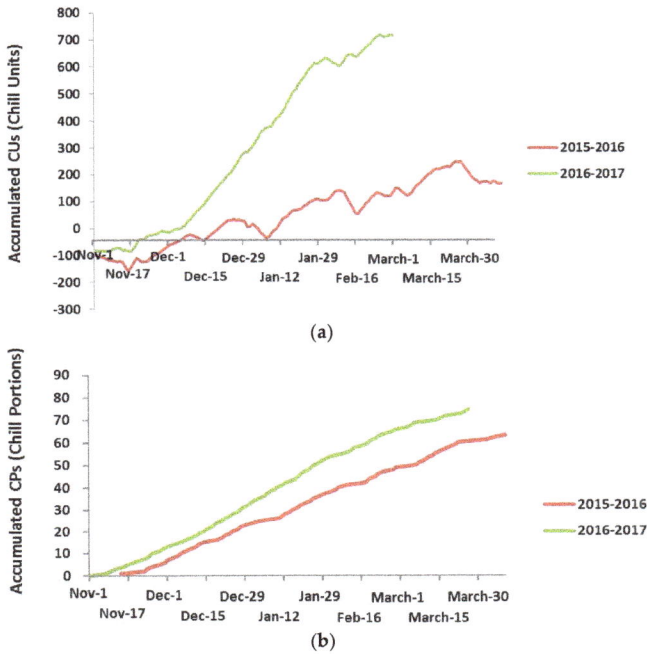

Figure 2. (a) Chilling accumulation in field conditions during the two seasons of study (2015–2016 and 2016–2017) calculated as chill units (CUs) according to the Richardson model; (b) calculated in chill portions (CPs) according to the dynamic model.

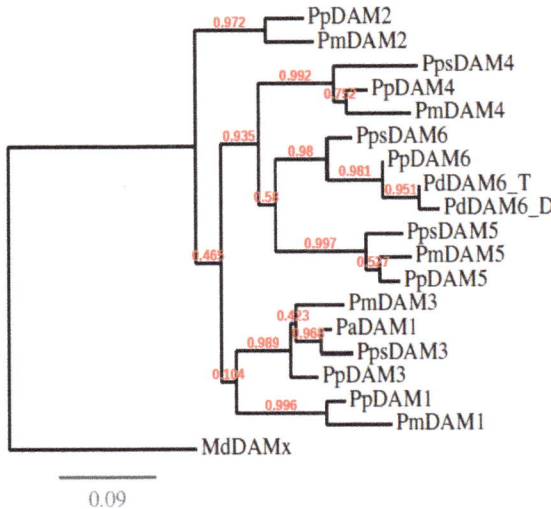

Figure 3. Phylogenetic tree showing relationships between homologous proteins to 'Desmayo Largueta' (PdDAM6_D) and 'Penta' (PdDAM6_P) sequence. *Prunus persica* (Pp), *Prunus mume* (Pm), *Prunus pseudocerasus* (Pps) and *Prunus avium* (Pa). aLRT (approximate likelihood ratio) values are indicated in red. *Malus domestica* DAM protein sequence (MdDAMx) was used as outgroup to root the phylogenetic tree.

3.3. Expression Analysis

Expression analysis of *PdDAM6* showed a progressive decrease in mRNA levels until the dormancy period was completely overcome, in both 'Desmayo Largueta' and 'Penta' cultivar samples (Figure 4).

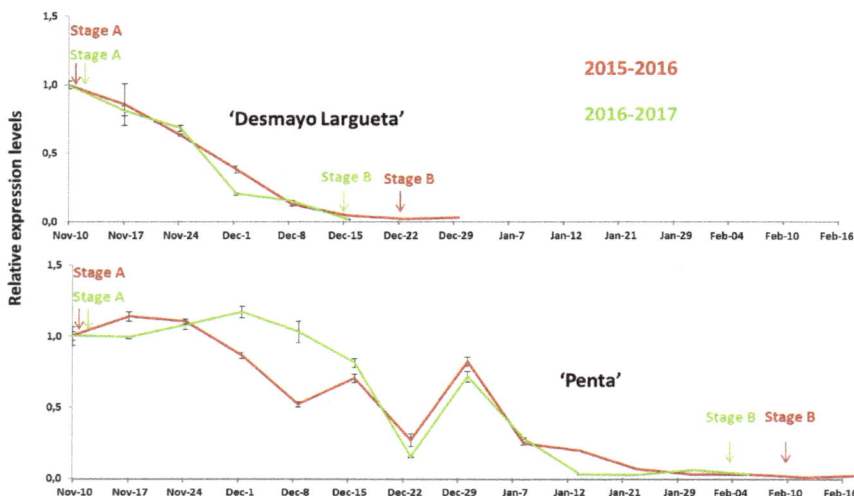

Figure 4. Relative gene expression of *PdDAM6* gene evaluated by qPCR (quantitative Polymerase Chain Reaction) 'Desmayo Largueta' and 'Penta' almond cultivars during the seasons 2015–2016 and 2016–2017. Standard deviations are indicated with vertical bars.

4. Discussion

As shown by Prudencio et al. [6], the estimation of chill accumulation under different climatic conditions showed that the dynamic model presents less variation than the Richardson model. As expected, the chilling requirements of the almond cultivars were related to their flowering time. However, in general, these values were lower than in previous evaluations performed by our group [3,8] mainly in the case of the warmer year.

The first full-length cDNA from the DAM gene family was obtained for almond. The clone designated *PdDAM6* (*Prunus dulcis DAM6*) was obtained from 'Desmayo Largueta' and 'Penta' almond cultivars, which display different phenotypes regarding chilling requirement and flowering time. Phylogenetic and expression analysis was performed to further characterize the sequences and to study the biological role of DAM proteins during flower bud dormancy progression in almond.

Our results clearly indicated that the level of expression of *DAM6* in both almond cultivars with different chilling requirements and flowering time decreased concomitantly with chill accumulation and dormancy progression, although for the late cultivar 'Penta', a relative increase was observed prior to dormancy release. These results supported that obtained by Leida et al. [14] and Jiménez et al. [15], highlighting the role of this gene in flower bud dormancy maintenance. In addition, a down-regulation of *DORMANCY-ASSOCIATED MADS-box6* has been observed in Japanese apricot [17,18] during dormancy release.

Monitoring bud transition from endodormancy to ecodormancy should be of great interest in terms of the use and optimization of biostimulants to promote flowering in fruit tree species [27,28] in the present climate change and warming context. The moment of application of these biostimulants is critical for success and depends on the endodormancy stage of the bud and its transition to ecodormancy [29] or the forcing strategies [30]. Treatments with these biostimulants should be applied at the optimum time for breaking bud dormancy, as they can be null or even toxic depending on the stage of the bud [29]. Monitoring almond flower bud dormancy through *DAM* expression could be used to determione the suitable moment to apply these biostimulants.

5. Conclusions

The estimation of chill accumulation using different models showed that the 2015–2016 season was warmer than the 2016–2017 season, and this was reflected in the dormancy release date of the cultivars. This illustrates the risk of growing extra-late cultivars in warm-winter areas, as production could be negatively affected if chilling requirement is not satisfied. The endodormancy to ecodormancy transition involves a transcriptional reprogramming in which genes acting on dormancy maintenance would be downregulated. This seems to be the case of *PdDAM6* for almond.

Author Contributions: A.S.P. and P.M.-G. participated in the design and coordination of the study. F.D. and P.M.-G. collaborated in the fieldwork. A.S.P. carried out the qPCR and cloning protocols. A.S.P., F.D. and P.M.-G. carried out data analysis. A.S.P., F.D. and P.M.-G. participated in the manuscript elaboration and discussion.

Funding: This study has been supported by Grants 19308/PI/14 and 19879/GERM/15 of the Seneca Foundation of the Region of Murcia and the Almond Breeding project of the Spanish Ministry of Economy and Competiveness.

Conflicts of Interest: The authors declare no conflicts of interest.

References

1. Lloret, A.; Badenes, M.L.; Ríos, G. Modulation of Dormancy and Growth Responses in Reproductive Buds of Temperate Trees. *Front. Plant Sci.* **2018**, *9*, 1368. [CrossRef] [PubMed]
2. Lang, B.A.; Early, J.D.; Martin, G.C.; Darnell, R.L. Endo-, para- and ecodormancy: Physiological terminology and classification for dormancy research. *Hort. Sci* **1987**, *22*, 371–377.
3. Egea, J.; Ortega, E.; Martínez-Gómez, P.; Dicenta, F. Chilling and heat requirements of almond cultivars for flowering. *Environ. Exp. Bot.* **2003**, *50*, 79–85. [CrossRef]
4. Campoy, J.A.; Ruiz, D.; Egea, J. Dormancy in temperate fruit trees in a global warming context: A review. *Sci. Hort.* **2011**, *130*, 357–372. [CrossRef]
5. Martínez-Gómez, P.; Prudencio, A.S.; Gradziel, T.M.; Dicenta, F. The delay of flowering time in almond: A review of the combined effect of adaptation, 406 mutation and breeding. *Euphytica* **2017**, *213*, 197. [CrossRef]
6. Prudencio, A.S.; Martínez-Gómez, P.; Dicenta, F. Evaluation of breaking dormancy, flowering and productivity of extra-late and ultra-late flowering almond cultivars during cold and warm seasons in South-East of Spain. *Sci. Hort.* **2018**, *235*, 39–46. [CrossRef]
7. Dicenta, F.; Sánchez-Pérez, P.; Batlle, I.; Martínez-Gómez, P. Late-blooming almond cultivar development. In *Almond: Botany, Production and Uses*; Rafael Socias i Company, Gradizel, T.M., Ed.; CABI: Boston, MA, USA, 2017.
8. Sánchez-Pérez, R.; Dicenta, F.; Martínez-Gómez, P. Inheritance of chilling and heat requirements for flowering in almond and QTL analysis. *Tree Genet. Gen.* **2012**, *8*, 379–389. [CrossRef]
9. Sánchez-Pérez, R.; Del Cueto, J.; Dicenta, F.; Martínez-Gómez, P. Recent advancements to study flowering time in almond and other *Prunus* species. *Front. Plant Sci.* **2014**, *5*, 334. [PubMed]
10. Bianchi, V.; Rubio, M.; Trainotti, L.; Verde, I.; Bonghi, C.; Martínez-Gómez, P. *Prunus* transcription factors: Breeding perspectives. *Front. Plant Sci.* **2015**, *6*, 443. [PubMed]
11. Bielenberg, D.G.; Wang, Y.; Fan, S.; Reighard, R.; Abbott, A.G. A Deletion Affecting Several Gene Candidates is Present in the Evergrowing Peach Mutant. *J. Hered.* **2004**, *95*, 436–444. [CrossRef] [PubMed]
12. Bielenberg, D.G.; Wang, Y.; Li, Z.; Zhebentyayeva, T.; Fan, S.; Reighard, G.L. Sequencing and annotation of the evergrowing locus in peach [*Prunus persica* (L.) Batsch] reveals a cluster of six MADS-box transcription factors as candidate genes for regulation of terminal bud formation. *Tree Genet. Gen.* **2008**, *4*, 495–507. [CrossRef]
13. Leida, C.; Terol, J.; Martí, G.; Agustí, M.; Llácer, G.; Badenes, M.L. Identification of genes associated with bud dormancy release in Prunus persica by suppression subtractive hybridization. *Tree Physiol.* **2010**, *30*, 655–666. [CrossRef] [PubMed]

14. Leida, C.; Conesa, A.; Llácer, G.; Badenes, M.L.; Ríos, G. Histone modifications and expression of DAM6 gene in peach are modulated during bud dormancy release in a cultivar-dependent manner. *New Phytol.* **2012**, *193*, 67–80. [CrossRef] [PubMed]

15. Jiménez, S.; Reighard, G.L.; Bielenberg, D.G. Gene expression of *DAM5* and *DAM6* is suppressed by chilling temperatures and inversely correlated with bud break rate. *Plant Mol. Biol.* **2010**, *73*, 157–167. [CrossRef] [PubMed]

16. Yamane, H.; Ooka, T.; Jotatsu, H.; Sasaki, R.; Tao, R. Expression analysis of *PpDAM5* and *PpDAM6* during flower bud development in peach (*Prunus persica*). *Sci. Hort.* **2011**, *129*, 844–848. [CrossRef]

17. Sasaki, R.; Yamane, H.; Ooka, T.; Jotatsu, H.; Kitamura, Y.; Akagi, T.; Tao, R. Functional and Expressional Analyses of *PmDAM* Genes Associated with Endodormancy in Japanese Apricot. *Plant Physiol.* **2011**, *157*, 485–497. [CrossRef] [PubMed]

18. Kitamura, Y.; Takeuchi, T.; Yamane, H.; Tao, R. Simultaneous down-regulation of *DORMANCY-ASSOCIATED MADS-box6* and *SOC1* during dormancy release in Japanese apricot (*Prunus mume*) flower buds. *J. Hort. Sci. Biotechnol.* **2016**, *5*, 476–482. [CrossRef]

19. Felipe, A.J. Phenological states of almond (In Italian). In Proceedings of the Third GREMPA Colloquium, Bari, Italy, 3–7 October 1977; pp. 101–103.

20. Richardson, E.A. ; S.D. Seeley, D.R.; Walker, J.L.; Anderson, M.; Ashcroft, G.L. Pheno-climatology of spring peach bud development. *Hort. Sci.* **1975**, *10*, 236–237.

21. Erez, A.; Couvillon, G.A.; Hendershott, C.H. The effect of cycle length on chilling negation by high temperatures in dormant peach leaf buds. *J. Am. Soc. Hort. Sci.* **1979**, *104*, 573–576.

22. Fishman, S.; Erez, A.; Couvillon, G.A. The temperature dependence of dormancy breaking in plants: Computer simulation of processes studied under controlled temperatures. *J. Theor. Biol.* **1987**, *126*, 309–321. [CrossRef]

23. Fishman, S.; Erez, A.; Couvillon, G.A. The temperature dependence of dormancy breaking in plants: Mathematical analysis of a two-step model involving a cooperative transition. *J. Theorl Biol.* **1987**, *124*, 473–483. [CrossRef]

24. Le Provost, G.; Herrera, R.; Paiva, J.A.; Chaumeil, P.; Salin, F.; Plomion, C. A micromethod for high throughput RNA extraction in forest trees. *Biol. Res.* **2007**, *40*, 291–297. [CrossRef] [PubMed]

25. Tong, Z.; Gao, Z.; Wang, F.; Zhou, J.; Zang, Z. Selection of reliable reference genes for gene expression studies in peach using real-time PCR. *BMC Mol. Biol.* **2009**, *10*, 71. [CrossRef] [PubMed]

26. Livak, K.J.; Schmittgen, T.D. Analysis of relative gene expression data using real-time quantitative PCR and the 2(-Delta Delta C(T)) Method. *Methods* **2001**, *25*, 402–408. [CrossRef] [PubMed]

27. Ionescu, I.A.; Moller, B.L.; Sánchez-Pérez, R. Chemical control of Flowering time. *J. Exp. Bot.* **2017**, *68*, 369–382. [CrossRef] [PubMed]

28. Ionescu, I.A.; López-Ortega, G.; Burow, M.; Bayo-Canha, A.; Junge, A.; Gericke, O.; Moller, B.L.; Sánchez-Pérez, R. Transcriptome and Metabolite Changes during Hydrogen Cyanimide-Induced Floral Bud break in Sweet Cherry. *Front. Plant Sci.* **2017**, *8*, 1233. [CrossRef] [PubMed]

29. Erez, A. Means to compensate for insufficient chilling to improve bloom and leafing. *Acta Hort.* **1995**, *395*, 81–95. [CrossRef]

30. Kauffman, H.; Blanke, M. Substitution of winter chilling by spring forcing for flowering using sweet cherry as model crop. *Sci. Hort.* **2018**, *244*, 75–81. [CrossRef]

horticulturae

MDPI

Review

Response of Mediterranean Ornamental Plants to Drought Stress

Stefania Toscano [1], Antonio Ferrante [2] and Daniela Romano [1],*

[1]   Department of Agriculture, Food and Environment (Di3A), Università degli Studi di Catania,
     Via Valdisavoia 5, 95123 Catania, Italy; stefania.toscano@unict.it
[2]   Department of Agricultural and Environmental Sciences, Università degli Studi di Milano, Via Celoria 2,
     1-20133 Milano, Italy; antonio.ferrante@unimi.it
*   Correspondence: dromano@unict.it; Tel.: +39-095-234-306

Received: 12 December 2018; Accepted: 3 January 2019; Published: 14 January 2019

Abstract: Ornamental plants use unique adaptive mechanisms to overcome the negative effects of drought stress. A large number of species grown in the Mediterranean area offer the opportunity to select some for ornamental purposes with the ability to adapt to drought conditions. The plants tolerant to drought stress show different adaptation mechanisms to overcome drought stress, including morphological, physiological, and biochemical modifications. These responses include increasing root/shoot ratio, growth reduction, leaf anatomy change, and reduction of leaf size and total leaf area to limit water loss and guarantee photosynthesis. In this review, the effect of drought stress on photosynthesis and chlorophyll *a* fluorescence is discussed. Recent information on the mechanisms of signal transduction and the development of drought tolerance in ornamental plants is provided. Finally, drought-induced oxidative stress is analyzed and discussed. The purpose of this review is to deepen our knowledge of how drought may modify the morphological and physiological characteristics of plants and reduce their aesthetic value—that is, the key parameter of assessment of ornamental plants.

Keywords: growth; gas exchange; chlorophyll fluorescence; oxidative stress; signal transduction; plant choice; green areas

1. Introduction

Drought stress strongly limits the growth of plants in Mediterranean regions. In the world, there are five Mediterranean-climate regions (i.e., areas surrounding the Mediterranean Sea, parts of western North America, parts of western and southern Australia, southwestern South Africa, and parts of central Chile) located between 32°–40° N and S of the Equator [1]. The Mediterranean climate is defined by precipitation and temperature, and it is characterized by a high seasonality summarized as hot and dry summers and cool and wet winters [2]. Despite the fact that these territories occupy less than 5% of the earth's surface, they harbor almost 20% of the world's vascular plant species [3]. The primary aspect that influences plant characteristics and natural vegetation is the extensive dry season. For this reason, plant growth and survival are endangered by long periods lacking rainfall and higher temperatures in the summer that impose more or less intense stress conditions [4]. The global climate changes that are occurring currently will worsen the availability of water, especially in arid and semi-arid environments. The availability of fresh and good quality water will decrease, especially in large cities [2,5]. This will entail difficulties in keeping green areas because the competition for water will be a critical issue.

For these reasons, great attention has recently been placed on the use and management of water to improve the sustainability of ornamental plant maintenance in semi-arid environments, such as the Mediterranean basin. Water scarcity led to the diffusion of techniques for creating green spaces that are able to save water (xeriscaping), favoring the use of species tolerant of water stress, which are native species like the carob tree, a species that is highly tolerant to high temperature and to low soil water efficiency [6]. This attention to water saving depends on the fact that even if the water in the urban environment is widely used for purposes other than irrigation (for example industrial and residential uses), "a landscape may serve as a visual indicator of water use to the public due to its visual exposure" [7]. The water saving can be maximized by utilizing different strategies such as making a suitable choice of ornamental plant species—one that has a high tolerance to drought stress without compromising the ornamental value and/or reducing the effects of drought stress through innovative cultivation methods.

Ornamental plants are not only species and/or cultivars that offer aesthetic pleasure, but they can also improve the environment and the quality of our lives [8]. Thus, ornamental plants can be used to restore disturbed landscapes, control erosion, reduce energy for climatization and water consumption, and improve the aesthetic quality of urban, peri-urban, and rural landscapes, as well as recreational areas, interiors, and commercial sites. In consideration of the many contexts in which plant species can be used for ornamental purposes, the number is very large. The wide number of the ornamental or potentially ornamental species increases the possibility of finding suitable genotypes that are able to cope with drought stress and that can be used for landscaping planning.

For landscaping, plant choice can be based on a very large number of species from a wide geographical range and with different functions [8]. Unlike in agriculture, performance of an amenity landscape is not measured with a quantifiable yield, but rather how well it meets the expectations of the user or the individual paying for installation and maintenance. These expectations include aesthetic appearance and/or utility such as shading, ground cover, and recreation [9]. Sometimes, in degraded environments, plant survival is the only purpose of cultivation. Furthermore, for ornamental plants used in landscaping, fast growth is not always desirable because the excessive shoot vigor often requires frequent pruning with higher management costs. To maintain a compact growth habit, ornamental plants may have to be pruned or treated with plant growth regulators [10]. A reduced quantity of water may have positive benefits on growth control, therefore moderate drought stress can be a useful tool to provide plants with compact habit and slower growth—both parameters required for easier landscape management [11]. Plant drought stress is difficult to study because the sensitivities and response times to water deficit vary among different plant species and are related to the intensity and length of the water stress. Plant response to drought stress involves the interaction of various physiological and biochemical parameters that can be exploited as markers for the identification of tolerant species [12].

2. Ornamental Plant Response to Drought Stress

2.1. Growth and Morpho-Anatomical Modification

Plant responses to drought are different and interconnected. Plant plasticity to drought stress adaptation varies within genera, species [13,14], and even cultivars. The main morphological changes under drought conditions are shoot and leaf growth reduction. These negatively affect the ornamental value and the visual appearance, which are particularly important key factors (from the ornamental point of view) that must be used along with markers for selecting tolerant genotypes [15]. Several experimental studies on ornamental plants showed that plant quality decreased in response to severe drought stress [16–18].

The effect of drought stress on plant growth and dry matter has been noticed in numerous ornamental species—for example, *Pistacia* [12], *Spiraea, Pittosporum* [19], *Bougainvillea* [20], *Callistemon* [21], *Laurus,* and *Thunbergia* [22] (Table 1). Since the photosynthetic pathways strongly influence the response to water stress, only the responses of C3 plants are presented in Table 1.

The reduction of leaf area is another typical response observed in plants subjected to water stress, as confirmed by several authors. Indeed, as reported by Toscano et al. [23], the total leaf area and the leaf number showed the widest variations in *Lantana* between control and severe deficit irrigation, while in *Ligustrum*, the differences were more marked for the total leaf area and not significant for the leaf number. The reduction of the leaf area is a consequence of a reduction in the leaf number [24] or the leaf size (unit leaf area) [22]. Thus, plants counteract the water limitation by reducing the transpiration area. One of the avoidance mechanisms that minimizes water loss when the stomata are closed is, in fact, the reduction of the canopy area. In callistemon plants, drought stress increases the root-to-shoot ratio, causing the reduction of aerial tissues rather than the roots [25–27]. This reduction also occurs when the plants are grown in pots, a frequent condition in the nursery phase.

Table 1. Major effects of drought stress on ornamental plants [1].

Species	Plant Habit	Treatments	Growth stage	Modified Parameter by Drought Stress	Ref.
Rudbeckia hirta, Callistephus chinensis, Althaea rosea, Malva sylvestris	forbs	4 levels of irrigation treatments: 25%, 50%, 75% and 100% of the reference evapotranspiration (ET0)	Seedling one month after transplanting	Plant fresh weight (−); SLA (−); Stomatal Conductance (−); Δ Canopy Temperature (+); water use efficiency index (WUEi) (+); water use efficiency biomass (WUEb) (+)	[28]
Periploca angustifolia	bushy-branched shrub	Full irrigation (FI), Water Deficit (WD), and Rehydrated (R)	11-month-old seedlings	Relative water content (RWC) (−); osmotic potential ($\psi\pi$) (−); water potential (ψw) (−); transpiration rate (−); net CO$_2$ assimilation rate (ACO$_2$) (−); stomatal conductance (g$_s$) (−); water use efficiency (WUE) (+); Proline (+); MDA (+); chlorophyll (a, b, total and a/b) and carotenoid content (−);	[29]
Pistacia lentiscus	bushy shrub	C = 100% water holding capacity; Moderate Water irrigation (MW, 60% of the control) and Severe Water deficit (SW, 40% of the C)	1-year-old seedlings	Dry weight (−), plant height (−) pre-dawn leaf water potential (Ψl) (−); RWC (−) in SW	[12]
Lantana camara, Ligustrum lucidum	bushy shrubs	C = container capacity, or irrigated at 100% of water container capacity (WCC); light deficit irrigation (LDI), irrigated at 75% of WCC; moderate deficit irrigation (MDI), irrigated at 50% of WCC; and severe deficit irrigation (SDI), irrigated at 25% of WCC.	Two-month-old rooted cuttings	Dry weight (−); leaf number (−); total leaf area (−); leaf thickness (−); photosynthesis (−); stomatal conductance (−); variable to maximal fluorescence (Fv/Fm) (−); water potential (−).	[23]
Bougainvillea buttiana 'Rosenka' and *B.* 'Lindleyana'	shrubby vines	C = substrate moisture close to container capacity and irrigation applied when 20% of the water was leached; deficit irrigation (DI), 25% of the amount of water supplied in C.	Two-year-old plants	Leaf, flower, total biomass dry weight, total leaf area (−); stomatal resistance (+); Ψl and Ψp (+); Stomatal length and width (−)	[20]
Spirea nipponica (S), *Pittosporum eugenioides* (P), *Viburnum nudum* (V)	bushy shrubs	4 irrigation levels (100, 70, 50, and 25% of container capacity) and Trinexapac-ethyl (TE) treatments (0.1, 0.2, and 0.3 L ha^{-1})	Plant heights 10 (S and V) and 40 cm (P)	Leaf number and area (−), plant dry weight and height (−), root dry weight (+). A, E, and gs (−). The application of 0.2 and 0.3 L ha^{-1} TE enhanced S, P and V tolerance to drought stress	[19]
Acacia tortilis subsp. *raddiana*	medium-sized tree	C = 80% of field capacity; Stress = withholding irrigation for 25 d.	6-week-old seedlings	Leaf number (−), dry mass (−), shoot length and total leaf area (−), water potential (−), stomatal conductance (−); transpiration rates (−); chlorophyll fluorescence (−) only when soil WC was < 40%, soluble sugars (+).	[30]
Viburnum opulus and *Photinia X fraseri* 'Red Robin'	bushy shrubs	C = 100% ET; Moderate Water Deficit plants (MWD) received 60% ET and Severe Water Deficit (SWD) received 30% ET.	Plants grown in pots (24 cm in diameter)	Water potentials (−); Pn and g$_s$ (−) in SWD in *P. x fraseri*; gs and leaf transpiration (Tr) (−) in *V. opulus*	[31]

Table 1. *Cont.*

Species	Plant Habit	Treatments	Growth stage	Modified Parameter by Drought Stress	Ref.
Callistemon laevis	bushy shrub	Control (0.8 dS m⁻¹, 100% water holding capacity), WD (0.8 dS m⁻¹, 50% of the amount of water supplied in control), saline (4.0 dS m⁻¹, same amount of water supplied as control) and saline water deficit (4.0 dS m⁻¹, 50% of the water supplied in the control).	2-year-old rooted cuttings	Total biomass (−); plant height (−); osmotic adjustment (−), leaf tissue elasticity (−)	[21]
Viburnum opulus L. and *Photinia x fraseri* 'Red Robin'	bushy shrubs	Control with 600 mL.day⁻¹ (C), moderate WD (MWD) 66% of C and severe water deficit (SWD) received 33% of C.	One-year-old plants	Stem diameter (−); Modulus of elasticity (−) only in *Photinia*	[32]
B. glabra 'Sanderiana', *B. xbuttiana* 'Rosenka', *B.* 'Lindleyana'	shrubby vine	Three irrigation levels based on the daily water use 100% (C), 50% (MDI) or 25% (SDI)	Rooted cuttings	SDW (−), total DW (−), leaf number (−), leaf area (−), macronutrient concentration (−) in SDI; Stomatal resistance (+), leaf water potential (−), leaf osmotic potential. (−)	[33]
Nerium oleander	bushy shrub	C (field capacity); WD (withholding irrigation)	One-year-old plants	Stem elongation (−); Leaf FW (−); Leaf WC (−); Chl (a, b and total) (−); Proline (+); Glycine betaine (+); Total soluble sugar (+); Total phenolic compounds (+); Total flavonoids (+); ascorbate peroxidase (+); glutathione reductase (+).	[34]
Callistemon citrinus 'Firebrand'	bushy shrub	C (substrate moisture close to container capacity); moderate deficit irrigation (MDI) by applying 50% of the amount of C and severe deficit irrigation (SDI) by applying 25% of the C irrigation	2-year-old rooted cuttings	RGR (−) in MDI; R/S ratio (+); WUE (+); g_s in MDI and SDI (−); P_n/g_s ratios (+); Stem water potential (−); P_n (−) in SDI	[24]
Pelargonium x hortorum	forb	C (100% of water field capacity = WFC); sustainable deficit irrigation (SDI), irrigated at 75% WFC throughout the experiment; regulated deficit irrigation I (RDI I), irrigated at 75% throughout the experiment, except during the flowering phase when plants were irrigated at 100%; regulated deficit irrigation II (RDI II), irrigated at 100% throughout the experiment, except during the flowering phase when plants were irrigated at 75%.	Rooted cuttings (4- to 5-cm tall and with 6–7 leaves)	Height (−), Flowering (−) RDI II; SDW (−), Number of leaves (−); Total leaf area (−).	[35]
Eugenia uniflora 'Etna Fire', *P. x fraseri* 'Red Robin'	bushy shrubs	Well-watered (WW), moderate drought stress (MD, 75%), severe drought stress (35%, SD).	Three months old rooted cuttings	A, g_s and E (−); RWC (−); Fv/Fm (−); Proline and MDA (+) in Eugenia; MDA (+) in SD.	[36]

Table 1. *Cont.*

Species	Plant Habit	Treatments	Growth stage	Modified Parameter by Drought Stress	Ref.
Myrtus communis	bushy shrub	Control (C), 100% water holding capacity [leaching 15% (v/v) of the applied water;]; moderate water deficit; MWD, 60% of the C; severe water deficit; SWD,40% of the C.	Seedlings of 2-year-old	SDW (−); root dry weights (−), leaf numbers (−), Total leaf area (−), plant height (−) in SWD; plant height (−) in MDW. Root hydraulic resistance (+); leaf water potential pre-dawn (−); Pn (−).	[37]
Pelargonium x hortorum	forb	Control, C, container capacity; Moderate deficit irrigation, MDI, 60% of the C; Severe deficit irrigation, SDI 40% of C. After 2 months, all the plants were exposed to a recovery period of 15 days with the same irrigation regime applied to control plants, until the end of the experiment.	Rooted cuttings	SDW (−); leaf area (−); R/S ratio (+); Height (−); Width (−); gs (−); Pn (−).	[38]
Callistemon citrinus, Laurus nobilis, Pittosporum tobira, Thunbergia erecta	bushy shrubs	Two consecutive cycles of suspension/rewatering (S-R) compared with plants that were watered daily (C).	Six-month-old plants	SDW (−); R/S ratio (+); RWC (−); Leaf water potential (−), gs (−); Pn (−).	[22]
Passiflora alata, P. edulis, P. gibertii, P. setacea, P. cincinnata	climbing vines	Two soil water regimes: soil field capacity and interruption of irrigation until the stomatal closure and apparent wilting of the whole plant.	Six month after sowing	Height (−); Dry weight of leaves, branches, roots (−); gs (−); palisade parenchyma thickness (+); leaf limb and spongy parenchyma thicknesses (+); leaf stomatal diameter (+).	[39]

[1] C = control; ET = Evapotranspiration; WD = Water deficit; SDW = shoot dry weight; (−) reduction due to WD; (+) increase due to WD.

Increased root-to-shoot ratios are frequently observed in plants under drought conditions which reduces water consumption [40] and increases water absorption [41]. This parameter is also suggested as a screening factor for grading plants with different stress tolerances. In addition to the reduction of the leaf area during drought stress, the modification of the leaf size and the cuticle thickness are also observed.

In a study conducted by Toscano et al. [23] on two ornamental shrubs (*Lantana* and *Ligustrum*) in a Mediterranean area, the analysis of leaf anatomical traits allowed the identification of the different strategies used during water stress conditions. During severe deficit irrigation, *Lantana* plants increased the spongy tissue rather than the palisade tissue; this anatomical modification facilitated the diffusion of CO_2 toward the fixation sites in order to increase the concentration gradient between internal air space and the atmosphere, thus enhancing the competition among cells for CO_2 and light [42]. In both species, an increase in the thickness of the spongy tissue and the palisade tissue was observed. The reduction of the specific leaf area could be a way to improve water use efficiency (WUE). In fact, thicker leaves usually have higher concentrations of chlorophylls and proteins per unit leaf area and thus have greater photosynthetic capacities per unit leaf area than thinner leaves [43].

The leaf anatomical modifications are species-specific. Thus, in *Polygala* and *Viburnum* plants subjected to four levels of irrigation treatments through the use of dielectric sensors (EC 5TE, Decagon Devices, Pullman, Washington, USA) to maintain the substrate water content (WS) equal to 10% (WC10%), 20% (WC20%), 30% (WC30%), and 40% (WC40% = control) of the pot volume, the leaf anatomical modifications were linked to spongy tissue in *Polygala* and palisade tissue in *Viburnum* (Figure 1).

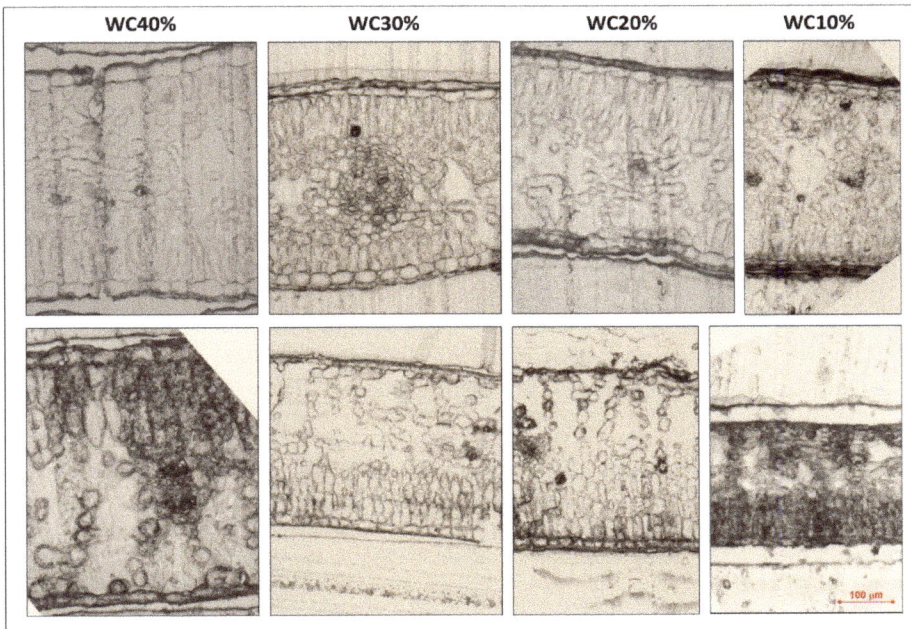

Figure 1. Light microscopy of blade cross-sections in *Polygala* (above) and *Viburnum* (below) at different water regimes (source: Toscano et al., unpublished data).

Acquiring greater knowledge of the morphological, physiological, and biochemical responses of the species in adverse environmental conditions, whether they are occasional, temporary, or long-term, allows us to choose the correct ornamental plants in relation to the interested area.

This information is useful in identifying the mechanisms of the adaptation of plants to adverse conditions such as drought stress [44], allowing us to select the most suitable species without compromising their aesthetic value.

2.2. Physiological Parameters

2.2.1. Leaf Gas Exchange

The main consequences of drought stress in plants are stomatal closure, reduction of gas exchange, the slowing down of photosynthetic activity, and the death of the plant [45,46]. Drought stress conditions mainly affect the photosynthetic system and ratio. In particular, they compromise the elements that are involved in the process, such as the electron transport to the thylakoids, the carbon cycle, and the stomatal control of CO_2 supply. Different published papers demonstrated that the reduction of photosynthetic activity is related to the mechanisms of stomatal conductance [47–50]. In fact, the first response of plants to water stress is stomatal closure and the subsequent reduction of the assimilation of the photosynthetic carbon necessary for the photosynthetic activity. As a consequence of the stomatal closure, there is not only a reduction in water loss, but also a reduction in nutrient uptake, consequently altering the metabolic pathways [51]. During drought stress, most species show a reduction in photosynthetic activity and a fast stomatal closure in relation to water potential adjustment [52,53]. The reduction in growth is also related to the reduction in the water potential of the leaves. Upon stomatal closure, a reduction in photosynthetic activity is achieved, which in turn leads to a decrease in plant growth and production [54,55]. The levels of carbon dioxide inside the stomatal chamber, and therefore in the cells, decrease, causing a reduction in photosynthesis. A decrease in the rate of CO_2 fixation is also observed and is associated with a reduction in the stomatal opening [56].

Under drought stress conditions, high conductivity ratios (A_N)/stomatal conductance (g_s) (also expressed as intrinsic WUE) indicate that leaves (the chloroplasts in particular)—even if there is an immediate stomatal closure—try to maintain high photosynthetic performance. As reported by Álvarez et al. [12], the decrease in g_s in *Pistacia lentiscus* subjected to drought stress limited water losses through transpiration control.

In order to estimate the tolerance to drought stress in plants, the transpiration ratio is essential. In fact, it has been observed that species that can retain a greater quantity of water and therefore lose less water through the stomata are more tolerant to drought. [57]. As reported by Galmes et al. [58], shrubs have a better ability to regulate transpiration compared to herbaceous plants.

2.2.2. Chlorophyll *a* Fluorescence

Under water stress conditions, one parameter that is commonly used to identify the presence of photosynthetic plant damage in plants is the measurement of chlorophyll *a* fluorescence. In fact, this parameter is very useful for analyzing the influence of environmental factors on the efficiency of the photosynthetic apparatus [59]. Down regulation of photosystem II (PSII) activity results in an imbalance between the generation and utilization of electrons, apparently resulting in changes in quantum yield [60]. The ratio variable to maximal fluorescence ratio (Fv/Fm) (i.e., the maximum primary photochemical efficiency of the PSII in a sample of leaves adapted in the dark) allows the evaluation of the efficiency of the PSII photosystem, indirectly measuring the physiological state of the plant [61]. Several authors have defined the Fv/Fm threshold values to indicate if a plant is more or less stressed. Values between 0.78-0.85 indicate that the plant is not stressed [62]. In a study conducted by Álvarez et al. [63] on *Callistemon* plants maintained at different levels of drought stress, the Fv/Fm values remained constant at 0.80. The drought stress was not compromised by the PSII. Therefore, the *Callistemon* is a species resistant to drought. Álvarez et al. [12] reported that in *Pistacia lentiscus* plants subjected to different levels of water stress (from May to October), low Fv/Fm values were found in stressed plants during the warmer months. At the end of the trial when the conditions were less stressful, the plants recovered from these values. This shows that the plants did not cause irreversible damage to the foliar tissues, indicating that PSII was not permanently damaged by stressful conditions. This affirms that the chloroplasts of Mediterranean species have different strategies during stress conditions for avoiding photo-inhibitory processes, such as the mechanism to consume the reducing power produced by the PSII [64,65].

2.3. Oxidative Stress

When photosynthetic activity is reduced and light excitation energy is in excess of that used or required by photosynthesis, over-excitation of the photosynthetic pigments in the antenna can occur, leading to the accumulation of reactive oxygen species (ROS) in chloroplasts [66]. During drought stress in plants, there are different biochemical changes. The main response is the accumulation of ROS, which causes the destruction of the cell membranes and results in oxidative damage to plants [67,68]. The plants, in order to oppose this accumulation, have developed many antioxidant activities and a series of secondary metabolites that counteract the generation of ROS and scavenge ROS once they are formed [69–71].

ROS are chemically active free radicals of oxygen. When unpaired electrons are present in the valence shell of these molecules, they become highly reactive and damage the cell structure and function. ROS production takes place within the compartments of different organelles, such as chloroplasts, mitochondria, and peroxisomes [60].

ROS include superoxide anion (O^{2-}), hydrogen peroxide (H_2O_2), hydroxyl radical (OH^-), singlet oxygen ($^1O_2$), and ozone (O_3). ROS are produced by plants continuously because they also have the role of cellular signaling, while excessive production involves oxidative stress [72].

Plants have mechanisms that protect them from the destructive action of oxidative reactions [73]. A mechanism put in place as a defense from stress relates to the production of antioxidant enzymes that protect the plants from ROS.

Garratt et al. [74] highlighted some enzymes among the main natural "detoxifiers" present in plants, such as superoxide dismutase (SOD; EC 1.15.1.1), catalase (CAT; EC 1.11.1.6), glutathione peroxidase (GPX; EC 1.11. 1.7), and ascorbate peroxidase (APX; EC 1.11.1.11). These enzymes are located in different compartments of the plant cells, while the CAT is instead located in the peroxisomes [75].

A type of ROS can be transformed into another type; for example, O_3 is decomposed into H_2O_2, O^{2-}, and $^1O^2$. The O^{2-} is also transformed spontaneously or enzymatically into H_2O_2 through SOD activity [76], which can react further with Fe^{2+} to form OH.

Controlling the production and action of ROS allows a better understanding of the effects of various abiotic stresses on plants. The study of protective mechanisms such as the antioxidant enzymes could allow the identification of processes that are the basis for the response of plants to stress.

When the plants are not stressed, the ROS level is kept low by the scavenger activity of the antioxidant enzymes. In the presence of abiotic or biotic stress (such as water, saline, or ozone stress), these balances are broken and there is an increase in the intracellular ROS levels. About 1%–3% of the oxygen that is consumed by plants leads to the formation of ROS [77,78]. The main changes that occur in plants are the increase in lipid peroxidation, protein degradation, DNA fragmentation, and finally cell death. All of this occurs because ROS are highly reactive [50]. Reacting with proteins and lipids, they modify structure, cellular metabolism, and, in particular, those that are linked to the photosynthetic process [79].

As a defense mechanism, the activity of these antioxidant enzymes increases under abiotic stress conditions such as drought [80–82], salinity [83,84], and ozone [85]. There are also non-enzymatic antioxidants: tocopherol, ascorbate, glutathione, phenols, alkaloids, flavonoids, and proline [60,72,86–91]. A decomposition product of poly-to-fatty acids of polyunsaturated fatty acids is malondialdehyde (MDA). It is considered a marker of membrane lipid peroxidation, which is an effect of oxidative damage. During the various drought stress conditions, some adapted species modify their antioxidant activities, increasing, for example, the activity of SOD and peroxidase (POD) [92]. SOD is the primary defense against ROS because it eliminates superoxide radicals. Specifically, it dismutates two O_2^-

radicals into H_2O_2 and O_2^-, which are precursors to other ROS and are generated in different subcellular compartments [93].

3. Mechanism of Signal Transduction and Development of Drought Tolerance

Drought stress is sensed by the roots of plants and the reduction of water availability slowly occurs depending on the soil physical properties. The limitation of water induces in plants several physiological, biochemical, and molecular changes that lead to increased plant tolerance (Figure 2). Since plants cannot escape from adverse weather conditions, survival depends on their ability to develop efficient adaptation strategies. The plant responses start from the activation of specific regulatory genes that lead to the modification of the physiology and the metabolism of the plants. Currently, transcriptional changes are widely studied in different species and under different drought stress conditions. Pioneer studies have been carried out on model plants, such as *Arabidopsis thaliana*, identifying the transcription profiles and transcription factors involved in responses to drought stress [94,95]. Among the different genes, dehydrin was found to be an indicator of the entire transcriptome response under drought stress. The increase in stress intensity induces the activation of genes associated with stress responses [96]. The most important genes involved belong to abscisic acid (ABA) perception and biosynthesis as well as the ethylene pathway. Among the transcription factors involved, the most important are abscisic acid-responsive element (ABRE), ABRE-binding (AREB) proteins, ABRE-binding factors (AREB/ABFs), drought-responsive *cis*-element binding protein/C-repeat-binding factor (DREB/CBF), ABF/AREB, NAC, WRKY transcription factors, Apetala 2 (AP2), and ethylene response elements [97,98]. The ABF/AREB are under ABA regulation involving SnRK2. These transcription factors are able to provide rapid gene activation under different abiotic stresses, including drought. Other transcription factors belong to the MYB family (such as MYB2 and MYC2) and are inducible by ABA [99]. Therefore, this plant hormone has a pivotal role under water stress in the activation of secondary gene networks, which leads to plant adaptation to stress. Mutants lacking ABA biosynthesis or action are very sensitive to drought stress [100].

The genes induced under drought encode for different proteins that are directly or indirectly involved in plant adaptation. Specific genes induced by water stress increase the accumulation of late embryogenesis abundant (LEA) proteins [101]. These proteins are accumulated in tissues under dehydration or desiccation, such as seeds. In plants, the LEA proteins are considered important in plant drought tolerance [102]. Water stress induces gene expression of membrane proteins. Among these, the most important are the aquaporins, i.e., the water channels.

At a biochemical level, plants increase the biosynthesis of osmolytes to lower the cell water potential and increase the water uptake ability of roots. These molecules are responsible for plant osmotic adaptation and include glycine, glycine betaine, proline, sugars, γ-aminobutyric acid, alcohols, sugar alcohols, trehalose, mannitol, polyamines, etc. [103–105]. The accumulation of these substances allows for the improvement of crop tolerance against drought stress, and the visual appearance of the plants does not change. Plants do not seem to be under stress conditions, but the biosynthesis of protectant molecules requires energy that is not exploited for the growth or the yield in agricultural crops. The energy used for the biosynthesis of osmolytes is also known as "fitness cost", which represents the energy costs for the plant to defend itself. The plants reduce photosynthetic activity, and ribulose-1,5-bisphosphate carboxylase/oxygenase (RUBISCO) efficiency declines with the increase in water reduction [106]. Since photosynthesis is a biochemical process that requires water, carbon dioxide, and light, the lack of water directly reduces photosynthesis. The quantum efficiency of PSII at the initial water stress transiently increases and then declines. The light received by the leaves must be dissipated to avoid photo-oxidative damage, and the energy dissipated can be estimated by chlorophyll *a* fluorescence. Gas exchange at the leaf level is regulated by stomatal opening. Under drought, water loss can be reduced by stomatal closure and a reduction in carbon dioxide concentration [107]. The reduction of light use can lead to an excess of excitation energy in leaves with ROS accumulation [108]. The increase in radicals stimulates the plant to activate the antioxidant systems, such as the enzymes involved in the detoxification of cells. The most important enzymes are SOD, CAT, APX, POD, glutathione reductase (GR), and monodehydroascorbate reductase (MDHAR). These enzymes are able to reduce the ROS accumulation and enhance plant tolerance to drought [70]. Drought stress is a common stress in plants grown in the Mediterranean area, and several ornamental shrubs subjected to water availability increase the activity of these enzymes [36].

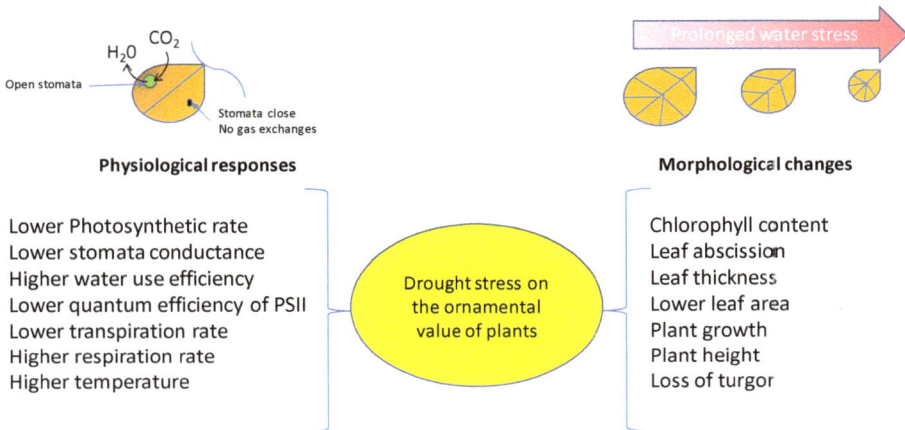

Figure 2. Physiological and morphological changes of plants exposed to reduced water availability. The magnitude of changes depends on the intensity of the stress.

Reduced photosynthetic activity also affects sugar concentration since respiration under drought increases because the plant temperature increases [109]. Plants under normal conditions are able to maintain the leaf temperature in the optimal range for photosynthesis by their thermoregulation ability, which is due to the evaporation of water at the leaf level through transpiration. The water passing from the aqueous state to the gas absorbs the heat from the plants and lowers the temperature. Under drought conditions, the closure of the stomata reduces transpiration and leads to a temperature increase, inducing a higher respiration rate. The lower photosynthesis and the higher respiration rate collectively reduce plant growth [110]. The reduction of plant growth in ornamental plants under water stress has been reported in several species, such as *Eugenia uniflora, Passiflora incarnata, and Photinia x* fraseri [36,111]. Ornamental plants can adopt different strategies under water stress. The study of plant responses to drought can be simulated by reducing water availability. In a study focused on drought responses, it was found that *Penstemon barbatus* was able to counteract drought by increasing root biomass and reducing stomatal conductance [112]. The gas exchange parameters, such as photosynthesis and stomatal conductance, can be considered good parameters for ornamental plant selection for tolerance to drought stress.

ABA is one of the most important plant hormones because it can regulate stomata opening in relation to potassium ions in guard cells [113]. An increase in ABA is crucial for reducing water loss through the stomata. Exogenous applications of ABA demonstrated that treated plants have a higher tolerance to drought. Another plant hormone that is induced by water stress is ethylene. It is also known as a senescence hormone because it is involved in leaf and flower senescence. Several ornamental plants are sensitive to ethylene, and it causes leaf abscission and yellowing, and petal rolling or desiccation [114]. Therefore, water stress can be detrimental for the ornamental plants used in the garden or other urban or peri-urban areas. Ethylene can be produced from endothermic engines. Therefore, in urban areas, plants exposed to ethylene and drought stress accelerate their senescence. Another important plant hormone that can have a positive role in the mitigation of drought stress is represented by the cytokinins. It has been demonstrated that *Arabidopsis* plants overexpressing genes involved in cytokinin biosynthesis showed higher drought stress tolerance [115]. These plant hormones have a preferential site of biosynthesis in roots, and drought stress seems to reduce the concentration of cytokinins with an increase in root growth [116]. The increase in root biomass is considered a first response of the plant to drought stress. However, the application of some plant growth promoting bacteria (PGPB) also induced drought tolerance by increasing their cytokinin concentration and ABA [117].

Therefore, plant adaptation to drought stress is due to plant hormone equilibrium, and the plant responses are consequences of the cross-talk among them [118].

4. Effects of Drought Stress on the Ornamental Value of Plants

Plants under water stress modify their morphology and physiology to survive under stressful environments. These changes can also have a direct effect on the visual appearance and subsequently the ornamental value of the plants. Morphological changes can be observed on the leaves and the plant habit. The most common changes that are observed are leaves that are smaller and have different orientations on the branches. Ornamental plants used in drought-prone environments must be able to adapt to the utilization area, such as private gardens or urban or peri-urban areas without irrigation systems. At nursery levels, the selection for drought environment can be carried out by considering the size and architecture of the roots, which can explore a wide volume of soil. Unfortunately, evaluation of root systems is not easy to perform.

In nursery cultivation, the generalized use of pots, often of small volume, cause root restriction effects. Yong et al. [119] analyzed the influence of substrate volume reduction on cotton plants under conditions where water and nitrogen supplies were not limited. The root-restriction lowered the rate of photosynthesis due to lower stomatal conductance. Root restriction increased the shoot-to-root ratios and reduced the total whole-plant leaf area by 20%.

The critical step for many ornamental plants is transplanting. Therefore, the hardening of plants is important for xerophytic environments [120]. After transplanting, the survival of plants can be guaranteed from their ability to reduce water losses through transpiration and gas exchange. The adapted plants must reduce stomatal conductance, maintain their water balance, and have high WUE [121].

The effects of drought have direct impacts on the habit of plants, and the ornamental quality can be observed at the leaf level. Leaves can drop, change color, or show necrosis from the action of ethylene. Flower life and turnover are also affected in many ornamental plants. The presence of flowers on plants greatly enhances their visual appearance. Therefore, tolerant plants should be able to have a high number of flowers with longer lives because, under water deficits, the turnover of flowers is reduced [122]. Flower turnover or new flower production depends on plant energy availability. Under prolonged drought stress, reduced photosynthesis and fewer carbohydrates are available for flowering.

However, reduced growth can have positive effects for urban green areas and the maintenance of public and private gardens due to lower management costs. Reduced growth is particularly important with ornamental plants that are shaped by pruning. Slower growth contributes to a longer preservation of shape with delayed pruning activities.

5. Use of Different Tools in Mitigating Drought-Induced Damages

A solution to overcoming the problems associated with drought stress is making an appropriate plant choice. The response to drought varies greatly among the plants that can be used in landscaping. In green areas, often a combination of woody and herbaceous ornamental plants is used with various manufactured elements (generally referred to as 'hardscape') [123]. The plant choice can refer to a very large number of species in different environments that are able to assure different functions in the landscape [8]. Plant adaptability to drought stress changes within genera, species, or cultivars [13,14].

Where drought stress is frequent, the ornamental plant choice can favor plants that grow in desert areas (like xerophytes or succulents), which are especially capable of surviving water shortages.

Arbuscular mycorrhizal (AM) symbiosis can also increase host resistance to drought stress, although the effect is not always predictable. Since drought stress is frequent in drying soils, the AM influence on plant drought response can be the result of AM influence on salt stress. With this aim, Cho et al. [124] determined if the AM-induced effects on drought responses would be more accentuated when plants of similar sizes were exposed to drought in salinized soils, rather than only when drought was applied. In the trial, using two greenhouses, different water relations characteristics were measured in sorghum (*Sorghum bicolor*) plants colonized by *Glomus intraradices*, *Gigaspora margarita*, or a mixture of AM species during a sustained drought following exposure to salinity treatments (NaCl stress, osmotic stress via concentrated macronutrients, or soil leaching). The findings confirmed that AM fungi can alter the host response to drought, but they did not lend much support to the idea that AM induced salt resistance. The beneficial effects of AM were related to the improved ability of the roots to adsorb water by increasing the active root surface. The increase in the root adsorption ability was also due to gibberellin- and cytokinin-induced production by AM [125].

Direct and indirect positive roles of PGPB in plants under stress have been reported [126]. The positive effects of PGPB are through the activation of 1-aminocyclopropane-1-carboxylate deaminase enzyme that reduces ethylene production and increases auxin concentration in roots [127]. In recent years, there has been an increase in biostimulants used in agriculture and horticulture to enhance crop abiotic stress tolerance [128]. Alleviation of abiotic stress is perhaps the most frequently cited benefit of biostimulant formulations [129]. Biostimulants are derived from organic substances through different industrial processes. They can be composed of microorganisms such as fungi or bacteria [128] and help to improve plant abiotic adaptation by acting on the physiology and biochemistry of plants [130]. The cytokinin-producing bacteria under drought conditions are of relevant interest [131]. Some microbial inoculants known to have a positive effect on plant development can also help plants overcome or tolerate abiotic stress conditions. In ornamental plants, production can be improved by biostimulant application. Hibiscus (*Hibiscus* spp.) treated with commercial biostimulants showed an increase in gas exchange with higher photosynthetic activities [132]. In a pot experiment with bedding plants, a seaweed extract of *Ascophyllum nodosum* revealed positive effects on the growth and development of petunias (*Petunia* spp.), pansies (*Viola tricolor*), and cosmos (*Cosmos* spp.) exposed to drought [133]. Some reported positive effects of biostimulants are the induction of early flowering, a higher number of flowers, and higher biomass accumulation [134]. With the aim of evaluating the differences in the mechanisms involved in ornamental species' resistance to drought stress resulting from a regular suspension and recovery of the water supply, Toscano et al. [22] subjected plants of five ornamental shrubs (*Callistemon citrinus*, *Laurus nobilis*, *Pittosporum tobira*, *Thunbergia erecta*, and *Viburnum tinus* 'Lucidum') to two consecutive cycles of suspension/rewatering (S-R) and compared them with plants that were watered daily (C). The five species exhibited different responses to drought stress. At the end of the experimental period, S-R treatment had no effect on the dry weight of any species except *Pittosporum*. In *Pittosporum*, drought stress reduced total plant biomass by 19%. Drought stress induced alterations in shrubs, including decreases in shoot dry matter and increases in the root-to-shoot ratio, strongly affecting *Callistemon* and *Pittosporum*. All species adapted to water shortages using physiological mechanisms (RWC and water potential adjustment, stomatal closure, and reductions in photosynthesis). Following re-watering, the species fully recovered. Therefore, they can be considered as suitable for landscaping in the Mediterranean environment. However, *Laurus* and *Thunbergia* seemed to be less sensitive to drought stress than the other species.

Light drought stress can be adopted to control the growth of pot plants. Davies et al. [135] used deficit irrigation in comparison to conventional overhead irrigation in two crops of different canopy structure (*Cornus alba* and *Lonicera periclymenum*). In a subsequent experiment, *Forsythia* × *intermedia* was grown in two substrates with contrasting quantities of peat (60 and 100%). Deficit irrigation was found to be mainly effective in controlling vegetative growth when applied using overhead irrigation.

Similar results were achieved when drip irrigation was used. This comparable response suggests that deficit irrigation can be applied without precision drip irrigation. Scheduling two very different crops with respect to their water use and uptake potential, however, highlighted challenges in the application of appropriate deficits for very different crops under one system. Responses to deficit irrigation are more consistent where nursery management allows for scheduling of crops with very different architecture and water use under different regimes.

6. Conclusions and Future Prospective

The drought tolerance of ornamental plants widely varies with genotypes, environmental conditions, and soil or substrate characteristics. Landscape plants have similar mechanisms of drought tolerance to agricultural crops, but assessment of drought tolerance for these plants should be based primarily on aesthetic value rather than growth effects. Because of the wide number of plant species potentially available for ornamental purpose, it should be possible to choose genotypes suitable for drought environments.

Problems in research that occur are linked to: (i) the necessity to experimentally analyze a wide range of plant species to find those most suitable for specific sites; (ii) identifying parameters with simple measurements to discriminate tolerance to drought stress, and (iii) tailoring irrigation methods or plant management strategies to enable the chosen species to cope with water stress.

The study of the mechanism of plant response to drought stress and particularly of signal transduction and development of drought tolerance allow for the identification of suitable plants and management strategies for the cultivation or utilization of ornamental plants in drought-prone environments.

Author Contributions: D.R. and A.F. projected the design of the review. S.T. and D.R. wrote the introduction and the information related with growth and morpho-anatomical modification under drought stress; S.T. wrote all the information related with physiological parameters and oxidative stress; A.F. wrote all the information related with mechanism of signal transduction and development of drought tolerance. D.R. and S.T. wrote all the information related with effect of drought stress on ornamental value, the tool use for mitigating the drought. S.T. and D.R. made the Figure 1 and Table 1. AF made the Figure 2. D.R. and S.T. ordered all the references. All authors wrote the conclusions and revised and approved the manuscript.

Funding: This research received no external funding.

Conflicts of Interest: The authors declare no conflict of interest.

References

1. Rundel, P.W.; Arroyo, M.T.; Cowling, R.M.; Keeley, J.E.; Lamont, B.B.; Pausas, J.G.; Vargas, P. Fire and plant diversification in mediterranean-climate regions. *Front. Plant Sci.* **2018**, *9*, 851. [CrossRef] [PubMed]
2. Paz, S.; Negev, M.; Clermont, A.; Green, M.S. Health aspects of climate change in cities with mediterranean climate, and local adaptation plans. *Int. J. Environ. Res. Public Health* **2016**, *13*, 438. [CrossRef] [PubMed]
3. Cowling, R.M.; Rundel, P.W.; Lamont, B.B.; Arroyo, M.K.; Arianoutsou, M. Plant diversity in Mediterranean-climate regions. *Trends Ecol. Evol.* **1996**, *11*, 362–366. [CrossRef]
4. Medrano, H.; Flexas, J.; Galmés, J. Variability in water use efficiency at the leaf level among Mediterranean plants with different growth forms. *Plant Soil* **2009**, *317*, 17–29. [CrossRef]
5. WWAP (World Water Assessment Programme). *The United Nations World Water Development Report 2014: Water and Energy*; UNESCO: Paris, France, 2014.
6. Ouzounidou, G.; Vekiari, S.; Asfi, M.; Gork, M.G.; Sakcali, M.S.; Ozturk, M. Photosynthetic characteristics of carob tree (*Ceratonia siliqua* L.) and chemical composition of its fruit on diurnal and seasonal basis. *Pak. J. Bot.* **2012**, *44*, 1689–1695.
7. Thayer, R.L. Visual ecology: Revitalizing the esthetics of landscape architecture. *Landscape* **1976**, *20*, 37–43.
8. Savé, R. What is stress and how to deal with it in ornamental plants? *Acta Hort.* **2009**, *813*, 241–254. [CrossRef]
9. Kjelgren, R.; Rupp, L.; Kilgren, D. Water conservation in urban landscapes. *HortScience* **2000**, *35*, 1037–1040.

10. Cameron, R.W.F.; Wilkinson, S.; Davies, W.J.; Harrison Murray, R.S.; Dunstan, D.; Burgess, C. Regulation of plant growth in container-grown ornamentals through the use of controlled irrigation. *Acta Hortic.* **2004**, *630*, 305–312. [CrossRef]

11. Niu, G.; Rodriguez, D.S.; Aguiniga, L.; Mackay, W. Salinity tolerance of *Lupinus havardii* and *Lupinus texenis*. *HortScience* **2007**, *42*, 526–528.

12. Álvarez, S.; Rodríguez, P.; Broetto, F.; Sánchez-Blanco, M.J. Long term responses and adaptive strategies of *Pistacia lentiscus* under moderate and severe deficit irrigation and salinity: Osmotic and elastic adjustment, growth, ion uptake and photosynthetic activity. *Agric. Water Manag.* **2018**, *202*, 253–262. [CrossRef]

13. Sánchez-Blanco, M.J.; Rodriguez, P.; Morales, M.A.; Torrecillas, A. Comparative growth and water relations of *Cistus albidus* and *Cistus monspeliensis* plants during water deficit conditions and recovery. *Plant Sci.* **2002**, *162*, 107–113. [CrossRef]

14. Torrecillas, A.; Rodriguez, P.; Sánchez-Blanco, M.J. Comparison of growth, leaf water relations and gas exchange of *Cistus albidus* and *C. monspeliensis* plants irrigated with water of different NaCl salinity levels. *Sci. Hortic. (Amsterdam)* **2003**, *97*, 353–368. [CrossRef]

15. Farieri, E.; Toscano, S.; Ferrante, A.; Romano, D. Identification of ornamental shrubs tolerant to saline aerosol for coastal urban and peri-urban greening. *Urban For. Urban Green.* **2016**, *18*, 9–18. [CrossRef]

16. Hansen, C.W.; Petersen, K.K. Reduced nutrient and water availability to *Hibiscus rosa-sinensis* 'Cairo Red' as a method to regulate growth and improve post-production quality. *Eur. J. Hort. Sci* **2004**, *69*, 159–166. [CrossRef]

17. Silber, A.; Levi, M.; Cohen, M.; David, N.; Shtaynmetz, Y.; Assouline, S. Response of *Leucadendron* 'Safari Sunset' to regulated deficit irrigation: Effects of stress timing on growth and yield quality. *Agric. Water Manag.* **2007**, *87*, 162–170. [CrossRef]

18. Bernal, M.; Estiarte, M.; Peñuelas, J. Drought advances spring growth phenology of the Mediterranean shrub *Erica multiflora*. *Plant Biol.* **2011**, *13*, 252–257. [CrossRef]

19. Elansary, H.O.; Salem, M.Z.M. Morphological and physiological responses and drought resistance enhancement of ornamental shrubs by trinexapac-ethyl application. *Sci. Hortic.* **2015**, *189*, 1–11. [CrossRef]

20. Cirillo, C.; De Micco, V.; Rouphael, Y.; Balzano, A.; Caputo, R.; De Pascale, S. Morpho-anatomical and physiological traits of two *Bougainvillea* genotypes trained to two shapes under deficit irrigation. *Trees Struct. Funct.* **2017**, *31*, 173–187. [CrossRef]

21. Álvarez, S.; Sánchez-Blanco, M.J. Comparison of individual and combined effects of salinity and deficit irrigation on physiological, nutritional and ornamental aspects of tolerance in *Callistemon laevis* plants. *J. Plant Physiol.* **2015**, *185*, 65–74. [CrossRef]

22. Toscano, S.; Scuderi, D.; Giuffrida, F.; Romano, D. Responses of Mediterranean ornamental shrubs to drought stress and recovery. *Sci. Hortic.* **2014**, *178*, 145–153. [CrossRef]

23. Toscano, S.; Ferrante, A.; Tribulato, A.; Romano, D. Leaf physiological and anatomical responses of *Lantana* and *Ligustrum* species under different water availability. *Plant Physiol. Biochem.* **2018**, *127*, 380–392. [CrossRef]

24. Álvarez, S.; Sánchez-Blanco, M.J. Changes in growth rate, root morphology and water use efficiency of potted *Callistemon citrinus* plants in response to different levels of water deficit. *Sci. Hortic.* **2013**, *156*, 54–62. [CrossRef]

25. Bacelar, E.A.; Santos, D.L.; Moutinho-Pereira, J.M.; Lopes, J.I.; Gonçalves, B.C.; Ferreira, T.C.; Correia, C.M. Physiological behaviour, oxidative damage and antioxidative protection of olive trees grown under different irrigation regimes. *Plant Soil* **2007**, *292*, 1–12. [CrossRef]

26. Chylińsku, W.K.; Łukaszewska, A.J.; Kutnik, K. Drought response of two bedding plants. *Acta Physiol. Planta* **2007**, *29*, 399–406. [CrossRef]

27. Bacelar, E.A.; Correia, C.M.; Moutinho-Pereira, J.M.; Gonçalves, B.C.; Lopes, J.I.; Torres-Pereira, J.M. Sclerophylly and leaf anatomical traits of five field-grown olive cultivars growing under drought conditions. *Tree Physiol.* **2004**, *24*, 233–239. [CrossRef]

28. Rafi, Z.N.; Kazemi, F.; Tehranifar, A. Morpho-physiological and biochemical responses of four ornamental herbaceous species to water stress. *Acta Physiol. Planta* **2019**, *41*, 7. [CrossRef]

29. Dghim, F.; Abdellaoui, R.; Boukhris, M.; Neffati, M.; Chaieb, M. Physiological and biochemical changes in *Periploca angustifolia* plants under withholding irrigation and rewatering conditions. *S. Afr. J. Bot.* **2018**, *114*, 241–249. [CrossRef]

30. Kebbas, S.; Lutts, S.; Aid, F. Effect of drought stress on the photosynthesis of *Acacia tortilis* subsp. *raddiana at the young seedling stage*. *Photosynthetica* **2015**, *53*, 288–298. [CrossRef]

31. Ugolini, F.; Bussotti, F.; Raschi, A.; Tognetti, R.; Ennos, A.R. Physiological performance and biomass production of two ornamental shrub species under deficit irrigation. *Trees Struct. Funct.* **2015**, *29*, 407–422. [CrossRef]

32. Ugolini, F.; Tognetti, R.; Bussotti, F.; Raschi, A.; Ennos, A.R. Wood hydraulic and mechanical properties induced by low water availability on two ornamental species *Photinia×fraseri* var. Red Robin and *Viburnum opulus* L. *Urban For. Urban Green.* **2014**, *13*, 158–165. [CrossRef]

33. Cirillo, C.; Rouphael, Y.; Caputo, R.; Raimondi, G.; De Pascale, S. The influence of deficit irrigation on growth, ornamental quality, and water use efficiency of three potted *Bougainvillea* genotypes grown in two shapes. *HortScience* **2014**, *49*, 1284–1291.

34. Kumar, D.; Al Hassan, M.; Naranjo, M.A.; Agrawal, V.; Boscaiu, M.; Vicente, O. Effects of salinity and drought on growth, ionic relations, compatible solutes and activation of antioxidant systems in oleander (*Nerium oleander*). *PLoS ONE* **2017**, *12*, e0185017. [CrossRef]

35. Álvarez, S.; Bañón, S.; Sánchez-Blanco, M.J. Regulated deficit irrigation in different phenological stages of potted geranium plants: Water consumption, water relations and ornamental quality. *Acta Physiol. Plant.* **2013**, *35*, 1257–1267. [CrossRef]

36. Toscano, S.; Farieri, E.; Ferrante, A.; Romano, D. Physiological and biochemical responses in two ornamental shrubs to drought stress. *Front. Plant Sci.* **2016**, *7*, 645. [CrossRef]

37. Navarro, A.; Álvarez, S.; Castillo, M.; Bañón, S.; Sánchez-Blanco, M.J. Changes in tissue-water relations, photosynthetic activity, and growth of *Myrtus communis* plants in response to different conditions of water availability. *J. Hortic. Sci. Biotechnol.* **2009**, *84*, 541–547. [CrossRef]

38. Sánchez-Blanco, M.J.; Álvarez, S.; Navarro, A.; Bañón, S. Changes in leaf water relations, gas exchange, growth and flowering quality in potted geranium plants irrigated with different water regimes. *J. Plant Physiol.* **2009**, *166*, 467–476. [CrossRef]

39. Souza, P.U.; Lima, L.K.S.; Soares, T.L.; de Jesus, O.N.; Filho, M.A.C.; Girardi, E.A. Biometric, physiological and anatomical responses of *Passiflora* spp. to controlled water deficit. *Sci. Hortic.* **2018**, *229*, 77–90. [CrossRef]

40. Wu, F.; Bao, W.; Li, F.; Wu, N. Effects of drought stress and N supply on the growth, biomass partitioning and water-use efficiency of *Sophora davidii* seedlings. *Environ. Exp. Bot.* **2008**, *63*, 248–255. [CrossRef]

41. Smirnoff, N. Plant resistance to environmental stress. *Curr. Opin. Biotechnol.* **1998**, *9*, 214–219. [CrossRef]

42. Fraser, L.H.; Greenall, A.; Carlyle, C.; Turkington, R.; Ross Friedman, C. Adaptive phenotypic plasticity of *Pseudoroegneria spicata*: Response of stomatal density, leaf area and biomass to changes in water supply and increased temperature. *Ann. Bot.* **2009**, *103*, 769–775. [CrossRef] [PubMed]

43. Liu, F.; Stützel, H. Biomass partitioning, specific leaf area, and water use efficiency of vegetable amaranth (*Amaranthus* spp.) in response to drought stress. *Sci. Hortic.* **2004**, *102*, 15–27. [CrossRef]

44. Chaves, M.M.; Maroco, J.P.; Pereira, J.S. Understanding plant responses to drought from genes to the whole plant. *Funct. Plant Biol.* **2003**, *30*, 239–264. [CrossRef]

45. Campbell, D.R.; Wu, C.A.; Travers, S.E. Photosynthetic and growth responses of reciprocal hybrids to variation in water and nitrogen availability. *Am. J. Bot.* **2010**, *97*, 925–933. [CrossRef] [PubMed]

46. Hu, X.; Liu, R.; Li, Y.; Wang, W.; Tai, F.; Xue, R.; Li, C. Heat shock protein 70 regulates the abscisic acid-induced antioxidant response of maize to combined drought and heat stress. *J. Plant Growth Regul.* **2010**, *60*, 225–235. [CrossRef]

47. Ahmadi, U.; Baker, D.A. The effect of water stress on grain filling processes in wheat. *J. Agric. Sci.* **2001**, *136*, 257–269. [CrossRef]

48. Del Blanco, I.A.; Rajaram, S.; Kronstad, W.E.; Reynolds, M.P. Physiological performance of synthetic hexaploid wheat–derived populations. *Crop Sci.* **2000**, *40*, 1257–1263. [CrossRef]

49. Samarah, N.H.; Alqudah, A.M.; Amayreh, J.A.; McAndrews, G.M. The effect of late-terminal drought stress on yield components of four barley cultivars. *J. Agron. Crop Sci.* **2009**, *195*, 427–441. [CrossRef]

50. Anjum, S.; Xie, X.Y.; Wang, L.C.; Saleem, M.F.; Man, C.; Wang, L. Morphological, physiological and biochemical responses of plants to drought stress. *Afr. J. Agric. Res.* **2011**, *6*, 2026–2032. [CrossRef]

51. Xiong, L.; Zhu, J. Molecular and genetic aspects of plant responses to osmotic stress. *Plant Cell Environ.* **2002**, *25*, 131–139. [CrossRef] [PubMed]

52. Ludlow, M.M. Stress physiology of tropical pasture plants. *Trop. Grassl.* **1980**, *14*, 136–145.

53. Nilsen, E.; Orcutt, D. *The Physiology of Plants under Deficit. Abiotic Factors*; Willey: New York, NY, USA, 1996; Volume 689, ISBN-13: 978-0471031529.

54. Guerrier, G. Fluxes of Na$^+$, K$^+$ and Cl$^-$, and osmotic adjustment in *Lycopersicon pimpinellifolium* and *L. esculentum* during short- and long-term exposures to NaCl. *Physiol. Plant.* **1996**, *97*, 583–591. [CrossRef]

55. Munns, R. Comparative physiology of salt and water stress. *Plant Cell Environ.* **2002**, *25*, 239–250. [CrossRef] [PubMed]

56. Inan, G.; Zhang, Q.; Li, P.H.; Wang, Z.L.; Cao, Z.Y.; Zhang, H.; Zhang, C.Q.; Quist, T.M.; Goodwin, S.M.; Zhu, J.; et al. Salt cress: A halophyte and cryophyte Arabidopsis relative model system and its applicability to molecular genetic analyses of growth and development of extremophiles. *Plant Physiol.* **2004**, *135*, 1718–1737. [CrossRef] [PubMed]

57. Riaz, A.; Younis, A.; Taj, A.R.; Riaz, S. Effect of drought stress on growth and flowering of marigold (*Tagetes erecta* L.). *Pak. J. Bot.* **2013**, *45*, 123–131.

58. Galmés, J.; Medrano, H.; Flexas, J. Photosynthesis and photoinhibition in response to drought in a pubescent (var. minor) and a glabrous (var. palaui) variety of *Digitalis minor*. *Environ. Exp. Bot.* **2007**, *60*, 105–111. [CrossRef]

59. Lichtenthaler, H.K.; Rinderle, U. The role of chlorophyll fluorescence in the detection of stress conditions in plants. *Crit. Rev. Anal. Chem.* **1988**, *19*, S29–S85. [CrossRef]

60. Reddy, A.R.; Chiatanya, K.V.; Vivekanandan, M. Drought induced responses of photosynthesis and antioxidant metabolism in higher plants. *J. Plant Physiol.* **2004**, *161*, 1189–1202. [CrossRef]

61. Maxwell, K.; Johnson, G.N. Chlorophyll fluorescence—A practical guide. *J. Exp. Bot.* **2000**, *51*, 659–668. [CrossRef]

62. Demmig, B.; Björkman, O. Comparison of the effect of excessive light on chlorophyll fluorescence (77K) and photon yield of O$_2$ evolution in leaves of higher plants. *Planta* **1987**, *171*, 171–184. [CrossRef]

63. Álvarez, S.; Navarro, A.; Nicolás, E.; Sánchez-Blanco, M.J. Transpiration, photosynthetic responses, tissue water relations and dry mass partitioning in *Callistemon* plants during drought conditions. *Sci. Hortic.* **2011**, *129*, 306–312. [CrossRef]

64. Flexas, J.; Medrano, H. Energy dissipation in C3 plants under drought. *Funct. Plant Boil.* **2002**, *29*, 1209–1215. [CrossRef]

65. Demmig-Adams, B.; Adams, W.W. Photoprotection in an ecological context: The remarkable complexity of thermal energy dissipation. *New Phytol.* **2006**, *172*, 11–21. [CrossRef] [PubMed]

66. Munné-Bosch, S.; Alegre, L. Drought-induced changes in the redox state of α-tocopherol, ascorbate, and the diterpene carnosic acid in chloroplasts of Labiatae species differing in carnosic acid contents. *Plant Physiol.* **2003**, *131*, 1816–1825. [CrossRef]

67. Smirnoff, N. The role of active oxygen in the response of plants to water deficit and dessication. *New Phytol.* **1993**, *125*, 27–58. [CrossRef]

68. Schwanz, P.; Picon, C.; Vivin, P.; Dreyer, E.; Guehi, J.M.; Polle, A. Responses of antioxidative system to drought stress in pendunculata oak and maritime pine as modulated by elevated CO2. *Plant Physiol.* **1996**, *110*, 393–402. [CrossRef] [PubMed]

69. Noctor, G.; Foyer, C.H. Ascorbate glutathione: Keeping active oxygen under control. *Annu. Rev. Plant Physiol. Plant Mol. Biol.* **1998**, *49*, 249–279. [CrossRef]

70. Mittler, R. Oxidative stress, antioxidants and stress tolerance. *Trends Plant Sci.* **2002**, *7*, 405–410. [CrossRef]

71. Foyer, C.H.; Noctor, G. Oxidant and antioxidant signalling in plants: A re-evaluation of the concept of oxidative stress in a physiological context. *Plant Cell Environ.* **2005**, *8*, 1056–1071. [CrossRef]

72. Impa, S.M.; Nadaradjan, S.; Jagadish, S.V.K. Drought stress induced reactive oxygen species and anti-oxidants in plants. In *Abiotic Stress Responses in Plants*; Springer: New York, NY, USA, 2012; pp. 131–147.

73. Foyer, C.H.; Descourvieres, O.; Kunert, K.J. Protection against oxygen radicals: An important defence mechanism studied in transgenic plants. *Plant Cell Environ.* **1994**, *17*, 507–523. [CrossRef]

74. Garratt, L.C.; Janagoudar, B.S.; Lowe, K.C.; Anthony, P.; Power, J.B.; Davey, M.R. Salinity tolerance and antioxidant status in cotton cultures. *Free Radic. Biol. Med.* **2002**, *33*, 502–511. [CrossRef]

75. Cruz de Carvalho, R.; Catala, M.; Silva, J.M.D.; Branquinho, C.; Barreno, E. The impact of dehydration rate on the production and cellular location of reactive oxygen species in an aquatic moss. *Ann. Bot.* **2012**, *110*, 1007–1016. [CrossRef] [PubMed]

76. Foyer, C.H.; Lopez-Delgado, H.; Dat, J.F.; Scott, I.M. Hydrogen peroxide and glutathione-associated mechanisms of acclamatory stress tolerance and signaling. *Physiol. Plant.* **1997**, *100*, 241–254. [CrossRef]

77. Matamorous, M.A.; Dalton, D.A.; Ramos, J.; Clemente, M.R.; Rubio, M.C.; Becana, M. Biochemistry and molecular biology of antioxidants in the rhizobia-legume symbiosis. *Plant Physiol.* **2003**, *133*, 499–509. [CrossRef] [PubMed]

78. Bhattacharjee, S. Reactive oxygen species and oxidative burst: Roles in stress, senescence and signal transduction in plant. *Curr. Sci.* **2005**, *89*, 1113–1121.

79. Lawlor, D.W.; Tezara, W. Causes of decreased photosynthetic rate and metabolic capacity in water-deficient leaf cells: A critical evaluation of mechanisms and integration of processes. *Ann. Bot.* **2009**, *103*, 561–579. [CrossRef]

80. Sankar, B.; Jaleel, C.A.; Manivannan, P.; Kishorekumar, A.; Somasundaram, R.; Panneerselvam, R. Effect of paclobutrazol on water stress amelioration through antioxidants and free radical scavenging enzymes in *Arachis hypogaea* L. *Colloids Surf B Biointerfaces* **2007**, *60*, 229–235. [CrossRef] [PubMed]

81. Jaleel, C.A.; Sankar, B.; Murali, P.V.; Gomathinayagam, M.; Lakshmanan, G.M.A.; Panneerselvam, R. Water deficit stress effects on reactive oxygen metabolism in *Catharanthus roseus*: Impacts on ajmalicine accumulation. *Colloids Surf. B Biointerfaces* **2008**, *62*, 105–111. [CrossRef] [PubMed]

82. Manivannan, P.; Jaleel, C.A.; Somasundaram, R.; Panneerselvam, R. Osmoregulation and antioxidant metabolism in drought stressed *Helianthus annuus* under triadimefon drenching. *C. R. Biol.* **2008**, *331*, 418–425. [CrossRef]

83. Jaleel, C.A.; Gopi, R.; Manivannan, P.; Gomathinayagam, M.; Murali, P.V.; Panneerselvam, R. Soil applied propiconazole alleviates the impact of salinity on *Catharanthus roseus* by improving antioxidant status. *Pestic. Biochem. Phys.* **2008**, *90*, 135–139. [CrossRef]

84. Manivannan, P.; Jaleel, C.A.; Kishorekumar, A.; Sankar, B.; Somasundaram, R.; Panneerselvam, R. Protection of *Vigna unguiculata* (L.) Walp. plants from salt stress by paclobutrazol. *Colloids Surf. B Biointerfaces* **2008**, *61*, 315–318. [CrossRef] [PubMed]

85. Puckette, M.C.; Weng, H.; Mahalingam, R. Physiological and biochemical responses to acute ozone-induced oxidative stress in *Medicago truncatula*. *Plant Physiol. Biochem.* **2007**, *45*, 70–79. [CrossRef] [PubMed]

86. Chen, C.; Dickman, M.B. Proline suppresses apoptosis in the fungal pathogen *Colletotrichum trifolii*. *Proc. Natl. Acad. Sci. USA* **2005**, *102*, 3459–3464. [CrossRef] [PubMed]

87. Jaleel, C.A.; Riadh, K.; Gopi, R.; Manivannan, P.; Ines, J.; Al-Juburi, H.J.; Chang-Xing, Z.; Hong-Bo, S.; Panneerselvam, R. Antioxidant defense responses: Physiological plasticity in higher plants under abiotic constraints. *Acta Physiol. Plant.* **2009**, *31*, 427–436. [CrossRef]

88. Gong, H.; Zhu, X.; Chen, K.; Wang, S.; Zhang, C. Silicon alleviates oxidative damage of wheat plant in pot under drought. *Plant Sci.* **2005**, *169*, 313–321. [CrossRef]

89. Yin, Y.; Li, S.; Liao, W.; Lu, Q.; Wen, X.; Lu, C. Photosystem II photochemistry, photoinhibition, and the xanthophylls cycle in heat-stressed rice leaves. *J. Plant Physiol.* **2010**, *167*, 959–966. [CrossRef] [PubMed]

90. Gill, S.S.; Tuteja, N. Reactive oxygen species and antioxidant machinery in abiotic stress tolerance in crop plants. *Plant Physiol. Biochem.* **2010**, *48*, 909–930. [CrossRef]

91. Ahmad, P.; Jaleel, C.A.; Salem, M.A.; Nabi, G.; Sharma, S. Roles of enzymatic and non-enzymatic antioxidants in plants during abiotic stress. *Crit. Rev. Biotechnol.* **2010**, *30*, 161–175. [CrossRef]

92. Lima, L.H.C.; Návia, D.; Inglis, P.W.; De Oliveira, M.R.V. Survey of *Bemisia tabaci* (Gennadius) (Hemiptera: Aleyrodidae) biotypes in Brazil using RAPD markers. *Genet. Mol. Res.* **2000**, *23*, 781–785. [CrossRef]

93. Alscher, R.G.; Erturk, N.; Heath, L.S. Role of superoxide dismutases (SODs) in controlling oxidative stress in plants. *J. Exp. Biol.* **2002**, *53*, 1331–1341.

94. Matsui, A.; Ishida, J.; Morosawa, T.; Mochizuki, Y.; Kaminuma, E.; Endo, T.A.; Satou, M. Arabidopsis transcriptome analysis under drought, cold, high-salinity and ABA treatment conditions using a tiling array. *Plant Cell Physiol.* **2008**, *49*, 1135–1149. [CrossRef] [PubMed]

95. Wilkins, O.; Bräutigam, K.; Campbell, M.M. Time of day shapes Arabidopsis drought transcriptomes. *Plant J.* **2010**, *63*, 715–727. [CrossRef]

96. Tommasini, L.; Svensson, J.T.; Rodriguez, E.M.; Wahid, A.; Malatrasi, M.; Kato, K.; Wanamaker, S.; Resnik, J.; Close, T.J. Dehydrin gene expression provides an indicator of low temperature and drought stress: Transcriptome-based analysis of barley (*Hordeum vulgare* L.). *Funct. Integr. Genom.* **2008**, *8*, 387–405. [CrossRef] [PubMed]

97. Uno, Y.; Furihata, T.; Abe, H.; Yoshida, R.; Shinozaki, K.; Yamaguchi-Shinozaki, K. Arabidopsis basic leucine zipper transcription factors involved in an abscisic acid-dependent signal transduction pathway under drought and high-salinity conditions. *Proc. Natl. Acad. Sci. USA* **2000**, *97*, 11632–11637. [CrossRef] [PubMed]

98. Klay, I.; Gouia, S.; Liu, M.; Mila, I.; Khoudi, H.; Bernadac, A.; Bouzayen, M.; Pirrello, J. Ethylene Response Factors (ERF) are differentially regulated by different abiotic stress types in tomato plants. *Plant Sci.* **2018**, *274*, 137–145. [CrossRef] [PubMed]

99. Abe, H.; Urao, T.; Ito, T.; Seki, M.; Shinozaki, K.; Yamaguchi-Shinozaki, K. Arabidopsis AtMYC2 (bHLH) and AtMYB2 (MYB) function as transcriptional activators in abscisic acid signaling. *Plant Cell* **2003**, *15*, 63–78. [CrossRef] [PubMed]

100. Koornneef, M.; Leon-Kloosterziel, K.M.; Schwartz, S.H.; Zeevaart, J.A. The genetic and molecular dissection of abscisic acid biosynthesis and signal transduction in Arabidopsis. *Plant Physiol. Biochem.* **1998**, *36*, 83–89. [CrossRef]

101. Magwanga, R.O.; Lu, P.; Kirungu, J.N.; Lu, H.; Wang, X.; Cai, X.; Zhou, Z.; Zhang, Z.; Salih, H.; Wang, K.; et al. Characterization of the late embryogenesis abundant (LEA) proteins family and their role in drought stress tolerance in upland cotton. *BMC Genet.* **2018**, *19*, 6. [CrossRef]

102. Hundertmark, M.; Hincha, D.K. LEA (late embryogenesis abundant) proteins and their encoding genes in *Arabidopsis thaliana*. *BMC Genom.* **2008**, *9*, 118. [CrossRef]

103. Bray, E.A. Plant responses to water deficit. *Trends Plant Sci.* **1997**, *2*, 48–54. [CrossRef]

104. Yoshiba, Y.; Kiyosue, T.; Nakashima, K.; Yamaguchi-Shinozaki, K.; Shinozaki, K. Regulation of levels of proline as an osmolyte in plants under water stress. *Plant Cell Physiol.* **1997**, *38*, 1095–1102. [CrossRef]

105. Hosseini, M.S.; Samsampour, D.; Ebrahimi, M.; Abadía, J.; Khanahmadi, M. Effect of drought stress on growth parameters, osmolyte contents, antioxidant enzymes and glycyrrhizin synthesis in licorice (*Glycyrrhiza glabra* L.) grown in the field. *Phytochemistry* **2018**, *156*, 124–134. [CrossRef] [PubMed]

106. Zandalinas, S.I.; Mittler, R.; Balfagón, D.; Arbona, V.; Gómez-Cadenas, A. Plant adaptations to the combination of drought and high temperatures. *Physiol. Plant.* **2018**, *162*, 2–12. [CrossRef]

107. Flexas, J.; Carriquí, M.; Nadal, M. Gas exchange and hydraulics during drought in crops: Who drives whom? *J. Exp. Bot.* **2018**, *69*, 3791–3795. [CrossRef] [PubMed]

108. Foyer, C.H.; Shigeoka, S. Understanding oxidative stress and antioxidant functions to enhance photosynthesis. *Plant Physiol.* **2011**, *155*, 93–100. [CrossRef]

109. Mariani, L.; Ferrante, A. Agronomic management for enhancing plant tolerance to abiotic stresses-drought, salinity, hypoxia, and lodging. *Horticulturae* **2017**, *3*, 52. [CrossRef]

110. Niu, G.; Rodriguez, D.S.; Wang, Y.T. Impact of drought and temperature on growth and leaf gas exchange of six bedding plant species under greenhouse conditions. *HortScience* **2006**, *41*, 1408–1411.

111. García-Castro, A.; Volder, A.; Restrepo-Diaz, H.; Starman, T.W.; Lombardini, L. Evaluation of different drought stress regimens on growth, leaf gas exchange properties, and carboxylation activity in purple Passionflower plants. *J. Am. Soc. Hortic. Sci.* **2017**, *142*, 57–64. [CrossRef]

112. Zollinger, N.; Kjelgren, R.; Cerny-Koenig, T.; Kopp, K.; Koenig, R. Drought responses of six ornamental herbaceous perennials. *Sci. Hortic.* **2006**, *109*, 267–274. [CrossRef]

113. Schroeder, J.I.; Kwak, J.M.; Allen, G.J. Guard cell abscisic acid signalling and engineering drought hardiness in plants. *Nature* **2001**, *410*, 327–330. [CrossRef] [PubMed]

114. Reid, M.S. Ethylene and abscission. *HortScience* **1985**, *20*, 45–50.

115. Prerostova, S.; Dobrev, P.I.; Gaudinova, A.; Knirsch, V.; Körber, N.; Pieruschka, R.; Fiorani, F.; Brzobohatý, B.; Cerný, M.; Spichal, L.; et al. Cytokinins: Their impact on molecular and growth responses to drought stress and recovery in Arabidopsis. *Front. Plant Sci.* **2018**, *9*, 655. [CrossRef]

116. Werner, T.; Nehnevajova, E.; Köllmer, I.; Novák, O.; Strnad, M.; Krämer, U.; Schmülling, T. Root-specific reduction of cytokinin causes enhanced root growth, drought tolerance, and leaf mineral enrichment in Arabidopsis and tobacco. *Plant Cell* **2010**, *22*, 3905–3920. [CrossRef]

117. Liu, F.; Xing, S.; Ma, H.; Du, Z.; Ma, B. Cytokinin-producing, plant growth-promoting rhizobacteria that confer resistance to drought stress in Platycladus orientalis container seedlings. *Appl. Microbiol. Biotechnol.* **2013**, *97*, 9155–9164. [CrossRef] [PubMed]

118. Peleg, Z.; Blumwald, E. Hormone balance and abiotic stress tolerance in crop plants. *Curr. Opin. Plant Biol.* **2011**, *14*, 290–295. [CrossRef] [PubMed]

119. Yong, J.W.; Letham, D.S.; Wong, S.C.; Farquhar, G.D. Effects of root restriction on growth and associated cytokinin levels in cotton (*Gossypium hirsutum*). *Funct. Plant Biol.* **2010**, *37*, 974–984. [CrossRef]

120. Franco, J.A.; Martinéz-Sanchéz, J.J.; Fernández, J.A.; Bañón, S. Selection and nursery production of ornamental plants for landscaping and xerogardening in semi-arid and environments. *J. Hortic. Sci. Biotechnol.* **2006**, *81*, 3–17. [CrossRef]

121. Lenzi, A.; Pittas, L.; Martinelli, T.; Lombardi, P.; Tesi, R. Response to water stress of some oleander cultivars suitable for pot plant production. *Sci. Hortic.* **2009**, *122*, 426–431. [CrossRef]

122. Rafi, Z.N.; Kazemi, F.; Tehranifar, A. Effects of various irrigation regimes on water use efficiency and visual quality of some ornamental herbaceous plants in the field. *Agric. Water Manag.* **2019**, *212*, 78–87. [CrossRef]

123. Iles, J.K. The science and practice of stress reduction in managed landscapes. *Acta Hortic.* **2003**, *618*, 117–124. [CrossRef]

124. Cho, K.; Toler, H.; Lee, J.; Ownley, B.; Stutz, J.C.; Moore, J.L.; Augé, R.M. Mycorrhizal symbiosis and response of sorghum plants to combined drought and salinity stresses. *J. Plant Physiol.* **2006**, *163*, 517–528. [CrossRef] [PubMed]

125. Barea, J.M.; Azcón-Aguilar, C. Production of plant growth-regulating substances by the vesicular-arbuscular mycorrhizal fungus Glomus mosseae. *Appl. Environ. Microbiol.* **1982**, *43*, 810–813.

126. Wong, W.S.; Tan, S.N.; Ge, L.; Chen, X.; Yong, J.W.H. The importance of phytohormones and microbes in biofertilizers. In *Bacterial Metabolites in Sustainable Agroecosystem*; Springer: Cham, Switzerland, 2015; pp. 105–158. [CrossRef]

127. Glick, B.R. Bacteria with ACC deaminase can promote plant growth and help to feed the world. *Microbiol. Res.* **2014**, *169*, 30–39. [CrossRef]

128. du Jardin, P. Plant biostimulants: Definition, concept, main categories and regulation. *Sci. Hortic.* **2015**, *196*, 3–14. [CrossRef]

129. Yakhin, O.I.; Lubyanov, A.A.; Yakhin, I.A.; Brown, P.H. Biostimulants in plant science: A global perspective. *Front. Plant Sci.* **2017**, *7*, 2049. [CrossRef]

130. Toscano, S.; Romano, D.; Massa, D.; Bulgari, R.; Franzoni, G.; Ferrante, A. Biostimulant applications in low input horticultural cultivation systems. *Italus Hortus* **2018**, *25*, 27–36. [CrossRef]

131. Calvo, P.; Nelson, L.; Kloepper, J.W. Agricultural uses of plant biostimulants. *Plant Soil* **2014**, *383*, 3–41. [CrossRef]

132. Massa, D.; Lenzi, A.; Montoneri, E.; Ginepro, M.; Prisa, D.; Burchi, G. Plant response to biowaste soluble hydrolysates in hibiscus grown under limiting nutrient availability. *J. Plant Nutr.* **2018**, *41*, 396–409. [CrossRef]

133. Battacharyya, D.; Babgohari, M.Z.; Rathor, P.; Prithiviraj, B. Seaweed extracts as biostimulants in horticulture. *Sci. Hortic.* **2015**, *196*, 39–48. [CrossRef]

134. Vernieri, P.; Ferrante, A.; Borghesi, E.; Mugnai, S. I biostimolanti: Uno strumento per migliorare la qualità delle produzioni. *Fertil. Agrorum* **2006**, *1*, 17–22.

135. Davies, M.J.; Harrison-Murray, R.; Atkinson, C.J.; Grant, O.M. Application of deficit irrigation to container-grown hardy ornamental nursery stock via overhead irrigation, compared to drip irrigation. *Agric. Water Manag.* **2016**, *163*, 244–254. [CrossRef]

horticulturae

Review

Agronomic Management for Enhancing Plant Tolerance to Abiotic Stresses: High and Low Values of Temperature, Light Intensity, and Relative Humidity

Antonio Ferrante [1] and Luigi Mariani [1,2,*]

[1] Department of Agricultural and Environmental Sciences, Università degli Studi di Milano, via Celoria 2, 20133 Milan, Italy; antonio.ferrante@unimi.it
[2] Lombardy Museum of Agricultural History, via Celoria 2, 20133 Milan, Italy
* Correspondence: luigi.mariani@unimi.it; Tel.: +39-329-702-7077

Received: 1 July 2018; Accepted: 16 August 2018; Published: 24 August 2018

Abstract: Abiotic stresses have direct effects on plant growth and development. In agriculture, sub-optimal values of temperature, light intensity, and relative humidity can limit crop yield and reduce product quality. Temperature has a direct effect on whole plant metabolism, and low or high temperatures can reduce growth or induce crop damage. Solar radiation is the primary driver of crop production, but light intensity can also have negative effects, especially if concurrent with water stress and high temperature. Relative humidity also plays an important role by regulating transpiration and water balance of crops. In this review, the main effects of these abiotic stresses on crop performance are reported, and agronomic strategies used to avoid or mitigate the effects of these stresses are discussed.

Keywords: cold; heat; solar radiation; relative humidity; transpiration

---

1. Introduction

Abiotic stress is the result of the action of external environmental factors that affect growth, development, and reproduction of crops. In the first part of this review published last year, we analyzed stressful conditions due to drought, water excess, salinity, and lodging [1]. In this part, high-low values of solar radiation, temperature, and relative humidity will be considered as well as agronomic strategies that can be used for lowering the stressful conditions.

Crop yield is the result of the interaction of multiple factors such as genotype, agronomic management, and environmental conditions. Different genotypes have different yield capabilities, depending on their ability to adapt. Agricultural cropping systems are continuously evolving due to innovation in agronomic tools and to identification of high-performance cultivars coming from traditional or biotechnological genetic improvements. Nevertheless, the major causes of agricultural production losses are due to abiotic stresses such as low water availability, high salinity, high or low temperatures, hypoxia/anoxia, and nutrient deficiency.

Crops exposed to these abiotic stresses respond by activating defense mechanisms. Therefore, in an early stress stage, no visible symptoms are exhibited. The energy used by the crop to counteract or cope with the abiotic stresses is called "fitness cost", and this energy does not contribute to production. While the visual appearance of plants in an early stress stage does not change, the physiology can undergo deep changes, including the accumulation of bioactive compounds able to counteract the stress conditions.

Plants are able to perceive environmental stimuli and to adapt to different environments. However, the degree of tolerance and adaptability to abiotic stresses varies among species and varieties. The global weight of biotic stresses on yield losses was estimated to be 70% by Boyer [2], and 13–94%

by Farooq et al. [3]. A more detailed analysis of yield losses associated with some abiotic stresses was presented by Mariani and Ferrante [1].

In this paper, a review of agronomic strategies aimed to optimize the resilience of crops exposed to abiotic stresses due to sub-optimal values of solar radiation, temperature, and relative humidity is presented. It is, however, necessary to remember that agronomic strategies hereafter presented and discussed can be adopted only if sustainable, not only socially and environmentally but also economically. This is because agriculture is an economic activity that cannot be done without adequate remuneration of the production factors.

2. Solar Radiation as Resource and Limitation for Crops

The sun provides the energy that moves the climate system (ocean and atmosphere circulation), drives the water cycle and feeds the food chains by means of the photosynthetic process. Moreover, an energy balance is in place for which the Earth emits towards space the same quantity of energy received from the sun.

Some basic physical laws rule the energy fluxes at all scales. More specifically the emission of energy by all bodies is ruled by Planck's Law, while the Stefan-Boltzmann Law and Wien's Law state that the total emission and the wavelength of the maximum emission are a function of the temperature of a blackbody, respectively. Moreover, the radiative flux (I) intercepted by a surface is determined by the sine law $I = I_0 * \sin(\alpha)$ where I_0 is the maximum radiative flux intercepted by a surface perpendicular to it and α is the elevation angle (the angle of the sun above the surface) [4]. The cosine law explains, for example, the high quantity of energy received by south-facing slopes at mid-latitudes during spring and autumn when the sun at noon is still low in the sky.

The radiation that comes directly from the sun is defined as beam radiation. The scattered and reflected radiation that arrives at the earth's surface from all directions (reflected from other bodies, molecules, particles, droplets, etc.) is defined as diffuse radiation. The sum of the beam and diffuse components is defined as total (or global) solar radiation (GSR). The fraction of the GSR re-irradiated toward space is named albedo A. Moreover, the surface of the planet emits radiation toward the sky in the far infrared (LR_1) and in its turn, the sky emits radiation in the far infrared towards the surface (LR_2). Consequentially, net radiation for a given surface is expressed with the following equation:

$$R_n = GSR * (1 - A) - LR_1 + LR_2 \tag{1}$$

Solar radiation is a source of energy and information for plants [5]. The relation between plants and light takes place through a series of pigments, which can be classified into four broad categories, namely chlorophylls, carotenoids, anthocyanins and phytochromes. Photosynthesis is the set of two phases, namely luminous and dark phase. In relation to the dark phase of photosynthesis, crops can be classified as C3, C4, and Crassulacean Acid Metabolism (CAM) plants [6].

For C3 plants the whole process of photosynthesis takes place in the mesophyll cells and the first products of photosynthesis catalyzed by Rubisco are two molecules with 3 atoms of carbon (Calvin cycle). In the case of C4 plants, the mesophyll is the site of CO_2 absorption by phosphoenolpyruvate (PEP), a reaction that is catalyzed by the enzyme PEP-carboxylase, which unlike Rubisco, is very efficient—even at low CO_2 levels. From the evolutionary point of view, C4 plants appeared on our planet much later than C3 plants (between 25 and 32 million years ago, according to Osborne and Beerling [7]) and were the product of the adaptation to high temperatures and water shortages.

CAM plants show a nocturnal phase in which, with open stomata, cells of the mesophyll produce OAA starting from PEP, and then malic acid is produced from OAA and is stored in the vacuolar juice that becomes increasingly acidic. During the day, when the stomata are closed, the malic acid returns to the cytoplasm where CO_2 is released from it which enters the Calvin cycle.

The relevance of C3 and C4 plants for global food security is indicated by the fact that about the 70% of the global calories required by humans is fulfilled by three C3 crops (wheat (*Triticum aestivum* L.), rice (*Oryza sativa*) and soybean (*Glycine max* L.) and two C4 crops (maize (*Zea mays*), sorghum (*Sorghum bicolor*)). On the other hand, CAM plants have lower importance and the only species subject to extensive cultivation are the pineapple (*Ananas comosus* (L.) Merr.) and the Indian-fig (*Ficus benghalensis*). For the C3 species, the lower efficiency of the Calvin cycle causes light saturation to occur at values between 30 and 80,000 lux (ex: 30–50,000 lux for grape (*Vitis vinifera* L.), while in the C4 saturation occurs at higher values at 80,000 lux. Therefore, on very bright days, when 100,000 lux is exceeded, the C3 species can exploit only a limited part of the available light energy. As far as light radiation is concerned, the compensation point is defined as the level of radiation at which photosynthesis and respiration reach equal values. Radiation values that coincide with this point are not significantly different between C3 and C4. At the level of general morphology, it is observed that leaves in the shade are wider and thinner (in particular the palisade layer appears less developed than that of leaves in the sun) and often have a greater concentration of chlorophyll in the upper surface. Added to this are a series of characteristic effects of the lack of radiation, i.e., yellowing and abscission of the lower leaves, a lack of branching, excessive elongation of stems and shoots and low or no fertility.

Photosynthesis is a key determinant of yield and quality of crops which in their turn are the pillar to the economic sustainability of farm activity. Therefore, all aspects of agronomic management should focus on crop photosynthesis with the objective of maximizing it even under stressful conditions. This objective can be achieved by selecting the best crops (species and varieties) for the selected environment, combining them in suitable crop successions and optimizing management techniques.

Making good choices of species and varieties are essential to making optimal use of the available solar radiation, avoiding at the same time problems due to radiation excess and water limitation. From this point of view, environments characterized by high temperatures, high values of global solar radiation, and water limitation are in general more suited to C4 plants, because these plants appeared some millions of years ago as an adaptation to savannas and tropical grasslands [8]. An important exception to this general rule is provided by maize, which in the presence of severe water stress shows a decrease in leaf area index (LAI), intercepted photosynthetically active radiation (PAR), and Radiation Use Efficiency (RUE) [9]. These drawbacks are usually associated with reduced competitiveness with weeds and proterandry that creates sterility. Such phenomena are only partially compensated by the increase of water use efficiency (WUE) and, consequently, a relevant reduction in maize biomass production and harvest index is generally observed.

Possible solutions for environments characterized by high temperatures and high levels of solar radiation could be use of C4 species such as sorghum, millet and panicum (Poaceae spp.) and sugarcane (*Saccharum officinarum*) or tropical C3's such as rice, oil palm (*Elaeis guineensis*), common bean (*Phaseolu vulgaris*) and cassava (*Manihot esculenta*). Some mesothermal C3 crops such as wheat and grapevine have shown significant injuries to photosystem II due to high temperatures [10].

Environments characterized by high values of cloud coverage have a greater presence of diffuse radiation that penetrates deeper into the canopy reaching the lower leaf layers. This explains why in these environments there is an increase in the RUE, which partially compensates for the decrease of intercepted PAR.

Obviously, the contrasting effects of the abovementioned factors (temperature, global solar radiation, LAI, intercepted PAR, RUE, soil water content, WUE) could be fully analyzed only through a dynamic crop simulation model [11–13] with a suitable time step (daily or hourly). This could be driven by weather data and run for a number of years sufficient to describe the climate of the selected environment (about 30 years is prescribed by the World Meteorological Organization to describe the climate for a given site).

Agronomic techniques (sowing density, plant nutrition, irrigation, integrated pest management, etc.) can promote the achievement of optimal values of LAI (for crops like maize, rice, sorghum, wheat, tomato (*Solanum lycopersicum*), and soybean the ideal values are 4–6), and should be adopted to maximize the light intercepted by crop canopies. Moreover, plants should be distributed as regularly as possible in the surface unit. From this point of view, an analysis carried out for apple (*Malus domestica*) orchards with the same LAI showed that square plantings (same distance between rows and along rows—1:1 ratio) intercept up to 20% more light than 3:1 system (3 times distance between rows than along rows) for a range of densities between 3000 and 20,000 trees per ha [14]. Other experiments have shown 10% higher yields at a 12% lower tree density than more rectangular single rows [15] (Wertheim, 1985). Moreover, square plantings give better coloration of red fruit than 3:1 plantings due to a more uniform light distribution. Despite these advantages, square plantings are rarely applied on a commercial scale because such plantings need higher capital investment for equipment adapted for over-the-row spraying and transport at harvest that may not compensate for higher yields, particularly with small-scale orchards [14].

Another relevant factor for optimizing light interception is the orientation of rows. In this regard, regardless of any other consideration (e.g., specific guidelines imposed by the effects of slope on mechanization or the presence of strong dominant winds able to break down rows perpendicular to them), in mid latitudes a north–south orientation is optimal for most crops, as it gives the maximum amount of intercepted radiation. In the specific case of grapevines, the east–west orientation of rows is preferable as it guarantees a greater radiation intercepted in coincidence with maximum daily temperatures, which for specific varieties is particularly appreciated in the late phase of sugar accumulation (September–October).

Specific interventions like pruning or choice of species/varieties with a suitable inclination of leaves can be adopted to enhance the penetration of light into the canopy so that each foliar level can ideally receive the same quantity of light. By this point of view, dicotyledon canopies (with almost planar leaves) are farther from this ideal than graminaceous canopies with a suitable inclination of leaves. Furthermore, in the specific case of maize, current cultivars, unlike traditional ones, have fully erect leaves able to enhance penetration of light inside the canopy and homogenize its vertical profile. Light extinction in canopies can be effectively simulated by adopting Lambert-Beer's law as stated by Monsi and Saeki in 1953 [16].

The sunlight intercepted by canopies is a fundamental driver of crop water consumption as expressed by the Priestley-Taylor equation for calculation of reference crop evapotranspiration (ET0) on the basis of net radiation. The sunlight intercepted by a canopy is also directly correlated with water consumption expressed as percentage of reference crop needs. For vineyards with row training systems, Williams and Ayars [17] found that the area shaded by the canopy at noon expressed as the percentage of the surface allotted to a single plant (sh%) multiplied by the empirical coefficient 0.017 gives a reliable estimate of crop coefficient (kc). This latter value can be multiplied by reference crop evapotranspiration ET0 to obtain the maximum evapotranspiration (ETM).

Moreover, solar radiation is a source of information for living beings and a series of morphogenetic effects derive from it. In many plants, initiation of flowering is driven by perception of changes in day length (photoperiodism). Plants adapted to temperate climates (e.g., photoperiodic varieties of wheat) often perceive the lengthening days of spring as a signal to initiate reproduction. Such plants are known as long-day plants, while in short day species (e.g., many varieties of chrysanthemum (*Chrysanthemum indicum*) and poinsettia (*Euphorbia pulcherrima*)) flowering is induced with hours of light below the critical threshold. Moreover photo-indifferent crops (e.g., many fruit tree species) are not sensitive to the length of the day [18].

Light signals such as light quality and length of day are perceived by several types of photoreceptors including phytochromes and cryptochromes. More specifically, phytochromes detect the levels and ratio of red (R) and far-red (FR) light in the environment [17,18]. Phytochromes are blue photosynthetic pigments present in the leaves and sensitive to very low light intensity (even lower than $0.01 \ \mathrm{W \ m^{-2}}$), which explains the effects of full moon light on flowering of some plant species [19,20].

Photoperiodic effects give rise to precautions in moving a plant species from one latitudinal band to another or the agronomic practices useful to induce flowering in flower species in protected cultivations.

2.1. High Light Intensity Stress

Solar radiation has a fundamental importance for crop growth, yield, and quality in agricultural systems. Light intensity and duration cannot be modified in the open field. Therefore, plants must adapt to light stress, while they can be modulated and optimized in a greenhouse. More specifically, crops in the open field during summer must protect themselves from high light intensity which can damage leaves, young shoots or even the fruits. Plants may protect chlorophyll molecules by increasing the biosynthesis and the concentration of carotenoids. These antioxidant compounds act as shields avoiding photo-oxidation of chlorophyll from excessive light intensity (Figure 1). High light intensity also leads to the formation of reactive oxygen species (ROS), which increase photo-damage [21,22]. Leaf damage from high light conditions can be determined by monitoring the lipid peroxidation of leaf cell membranes and the functionality of photosystem II. The most dangerous radiation of the light spectrum that induces severe damage is UV-B radiation (280–320 nm) that has short wavelengths and high energy. The damage from UV-B radiation is especially notable on the vital macromolecules such DNA with negative effects on cellular processes. Light stress reduces the light use efficiency (LUE) and photosynthetic activity and can be observed in both the open field and in protected cultivation.

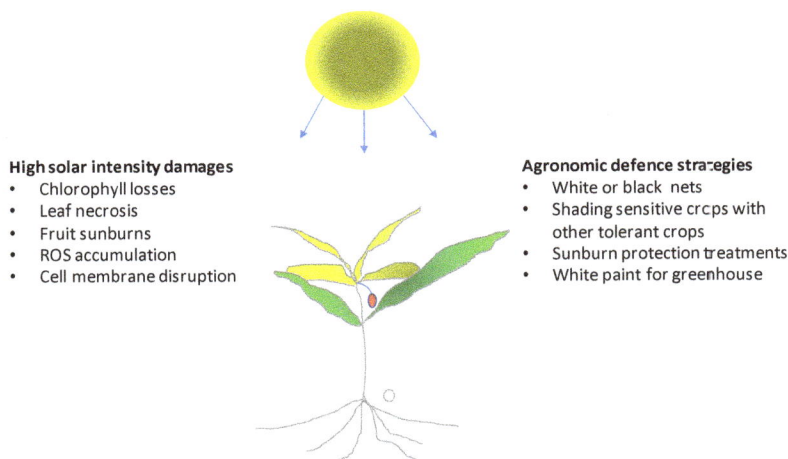

High solar intensity damages
- Chlorophyll losses
- Leaf necrosis
- Fruit sunburns
- ROS accumulation
- Cell membrane disruption

Agronomic defence strategies
- White or black nets
- Shading sensitive crops with other tolerant crops
- Sunburn protection treatments
- White paint for greenhouse

Figure 1. Effects of high light intensity on crops and agronomic strategies that can be adopted for increasing or avoiding crop or produce damage.

In the open field, light stress is particularly severe if associated with high temperature and drought. Agronomic strategies for reducing the negative effects of an excess of light include adequate irrigation systems, with support of sensors able to evaluate soil moisture and crop water requirements. Sufficient water availability can guarantee transpiration and, thus, thermoregulation by the evaporation of water at the leaf level which lowers leaf temperature.

In mixed crops, species with different light stress tolerances and heights can be cultivated together to provide shading of the most sensitive crop species. This strategy can be used to protect young seedlings of sensitive species against damage by strong solar radiation but can give protection against other stress factors like strong winds, low temperatures or salt. Once seedlings are sufficiently developed the crop used to protect them can be eliminated manually or by means of a selective herbicide treatment.

A high light intensity can reduce the yield and quality of young fruit directly exposed to sunlight. This is mostly important for fruit of species whose stomata no longer allow evapotranspiration with the approach of maturation, which makes them unable to thermoregulate fruit temperature. Therefore, fruit temperature shows an increase, especially with dark-colored fruit (e.g., grapevine berries of red varieties after veraison), when temperature in the presence of direct solar radiaton can exceed 45–50 °C. In these cases, damage can be direct (sunburn with cell membranes that lose their integrity) or indirect (slowing or interruption of the biosynthesis of compounds favorable to quality). Temperature of fruit can be measured directly by means of thermocouples or simulated by energy balance models [23].

Recently several products have become commercially available that can be sprayed on plants or fruit for avoiding sun damage. Kaolin containing compounds have been successfully applied to reduce sunburn in pomegranate [24]. Good results can be also obtained using shading nets. In apple trees, transparent and black nets were able to reduce sunburn incidence on fruit [25]. The nets reduced the direct sun light on fruit avoiding the excessive temperature of the exposed tissue and localized physiological disorders. Moreover, in vineyards, canopy management techniques (e.g., winter or green pruning) have protected clusters by covering them with a layer of leaves, which in many cases has enhanced berry quality [26].

In greenhouses during summer, high temperature associated with high light intensity can induce direct damage on plants. Therefore, to avoid the high temperature and light intensity, greenhouses are covered with shading nets or sprayed with white paint. As an alternative, white shading nets can also be placed inside the greenhouse. The aim of these strategies is to avoid excess heat in the greenhouse through the reduction of direct light. Water running along the roof and walls of the greenhouse in special interspaces also has potential application, since water reduces the radiation entering the greenhouse.

2.2. Low Light Intensity Stress

Some plant species are able to grow at low light conditions (termed shade plants) such as under vegetation or in areas with low solar radiation (e.g. valley bottoms, entering of caves). Most agricultural crops need high light intensity and are classified as sun plants. Shade plants have low light compensation and saturation points, while sun plants in contrast have high light compensation and saturation points. However, most plants can adapt to a range of low light conditions. The adaptation induces physiological and morphological changes in plants exposed to such conditions. In general, low light intensities induce stem elongation to overcome the shade conditions. Leaves of shaded plants increase their size and reduce their thickness and have a higher chlorophyll concentration. At a physiological level, the plants lower their light compensation point for balancing the reduced photosynthetic activity.

Low light conditions present in greenhouses during winter can be overcome by supplemental lighting using fluorescent lights, metal halide lights, high-pressure sodium lamps (HPS), or light emitting diodes (LEDs). The outputs of these lamps ideally must match the crop light utilization spectra. If the lamps have a higher emission in the regions of the leaf absorbance spectra, the LUE as well as the yield and quality are higher. LEDs can provide precise outputs and emission spectra (wavelengths) and can be readily adjusted based on the species requirements. Few studies have been performed on the effects of light intensity and quality on plant morphological and developmental processes [27]. Plants sense and translate environmental light signals, which interfere directly or indirectly with metabolism [28]. Light can modify auxin transport and gibberellin (GA) biosynthesis [29,30]. These changes can affect many aspects of plant development, including seed germination, stem elongation, and floral initiation. Light quality can modify the profile of carbon and nitrogen metabolites and those of organic acids and aromatic amino acids [27,31]. An alteration of the organic acids level, in particular of α-ketoglutarate, can affect amino acid biosynthesis, plant growth, and crop yield [27]. Supplemental lighting is often very expensive, and its application is only used for the most lucrative crops such as tomato or rose.

3. Thermal Resources and Limitations for Crops

3.1. Temperature and Agriculture

Typically, in agriculture we refer to surface air temperature, which is the temperature measured with a thermometer placed about 1.50–1.80 m above a soil with a regularly mowed lawn and protected by an anti-radiation screen that ensures a suitable circulation of air around the sensor. This measure is used to estimate thermal resources and limitations for cultivated and spontaneous plants.

The temperature that would be most useful to know is that of the organs of plants. The leaf temperature is mitigated by the transpiration process, which transfers to the atmosphere 2450 J g^{-1} via transpired water. Consequently, the leaf temperature of well-watered plants is close to that of the air. However, when soil water is insufficient, plants close their leaf stomata, and the leaf temperature rises to levels significantly higher than the air.

Another important variable is the temperature of the soil layer explored by roots. It can be measured at different depths, although usually for agricultural purposes it is only gauged at 10 cm in depth because the most superficial soil layer is also the only one that shows a significant daily cyclicity.

3.2. Temperature and Some Physical Presuppositions

Temperature values and variability in space and time obey physical presuppositions that are briefly summarized here. Hot and cold air behave as non-miscible fluids and cold air, being denser, flows downward along the sides of hills and mountains and gathers in valleys, depressions and concavities of the ground. Moreover, during the day, the sun warms the ground and this warms the air layers over it while, during the night, the ground cools radiating energy toward space and cools the layers of air over it. The strength of the ground radiative cooling towards space is directly proportional to the fraction of sky visible from it (sky view factor (SVF)), and it is at a maximum for a flat plain without surrounding obstacles (mountains, trees, walls, buildings, etc.) or for the tops of hills and mountains. These latter locations have a higher SVF than the bottom of the valleys, which also cool indirectly due to the cooler dense air that descends from higher elevations along the slopes with cold air accumulation in the lowlands (cold lake effect).

Above the cold lake there is a milder area called a thermal belt, which is an area of thermal optimum. Historically, villages of the European Alps and Italian Apennines were often built in the thermal belt mainly because the need for winter heating is lower.

The most favorable exposure, from the perspective of thermal resources, is the south face, while the north face is the least favored. In an intermediate position with respect to these two extremes are the sides facing west and east, the latter being thermally more favored because the eastward-side is the first to receive the sun in the morning, when light energy must warm the surfaces that are cold after the night and are often dew-covered (informally described as "the sun works on the cold") before the air is heated. When the sun illuminates the west facing slopes in the afternoon, it heats already warm slopes.

Maximum and minimum daily air temperatures are mainly determined by:

1. The energy balance of the underlying surface;
2. The short distance transport by downslope and upslope air movement;
3. The advection of air masses from more distant areas; for example, in Europe, the advection of subtropical air masses from the intertropics, Arctic air masses from the polar region and polar continental air masses from Siberia;
4. The rising of air from the surface resulting in cooling (convection);
5. The compressional effect typical of dynamic anticyclones where air masses are animated by a vertical descending motion of high pressure.

In addition, soil covered by vegetation heats up less than a bare soil during the day and cools less during the night, and soil covered with a sufficiently thick snow layer shows a constant temperature near the surface close to 0 °C, which also maintains some microbial activity.

3.3. Quantitative Approach to Temperature Resources and Limitations for Plants

Temperature is a primary driver for plant growth and development (G&D). Therefore, the objective of the grower is to optimize thermal resources and minimize limitations, to meet the business objectives (i.e., quantity and quality of crop production). Given specific values of daily or hourly temperature, to discriminate between resources and limitations, adequate G&D happen only in a specific thermal range delimited by the two reference values, the lower cardinal (LC) and the upper cardinal (UC) temperatures [32]. Furthermore, in this primary thermal range it is possible to detect a sub-optimal range (delimited by lower optimal (LO) and upper optimal (UO) temperatures) where G&D happens without thermal limitation. Below the LC and above the UC values, there are mortality thresholds for low (CL) and high (CH) temperatures [25].

Values of cardinal temperatures for various species are shown in Table 1. However, it should be noted that cardinal temperatures for each species and variety may vary with phenological stage. For wheat, the values of LC, LO, UO, and UC are 3.5, 20, 24, and 33 °C, respectively, from sowing to emergence; −1.5, 4, 6, and 15.7 °C for vernalization; 1.5, 9, 12, and >20 °C for terminal spikelet; and 9.5, 19, 23, 31 °C for anthesis and 9.2, 19, 22, 35.4 °C for caryopsis development until ripening. On the other hand, LC, LO, UO and UC for some general processes of wheat are −1, 21.5, 22.5, 24 °C, respectively, for leaf initiation; 3, 20, 21, >21 °C for shoot growth; and 2, 12.5, 20, >25 °C for root growth [27,33].

If T_m is the daily average temperature, daily thermal resources useful for G&D can be expressed as growing degree days (GDD) or normal heat hours (NHH). GDD are calculated by the following equation:

$$GDD = (T_m - LC) \tag{2}$$

where GDD = 0 for T_m < LC. Moreover, to evaluate the depleting effect of high temperatures on G&D, the following alternative truncated conditions are adopted: If T_m > UC then T_m = UC or T_m = UC − $(T_m - UC)$.

A more physiologically-based way to simulate thermal resources, taking into account the effect of sub-optimal or supra-optimal temperatures, is provided by the NHH method that converts hourly average temperatures to a normal value in the range 0–1 by means of a suitable response curve like that shown in Figure 2, growing from 0 to 1 between LC and LO, equal to 1 between LO and UO and linearly decreasing from 1 to 0 between UO and UC [34]. This method enables quantification of thermal limitation by low temperatures, when normal hours are useless because they are sub-optimal (LHH), and by high temperatures, when normal hours are useless because they are supra-optimal (HHH), respectively expressed as a complement to 1 for the NHH value.

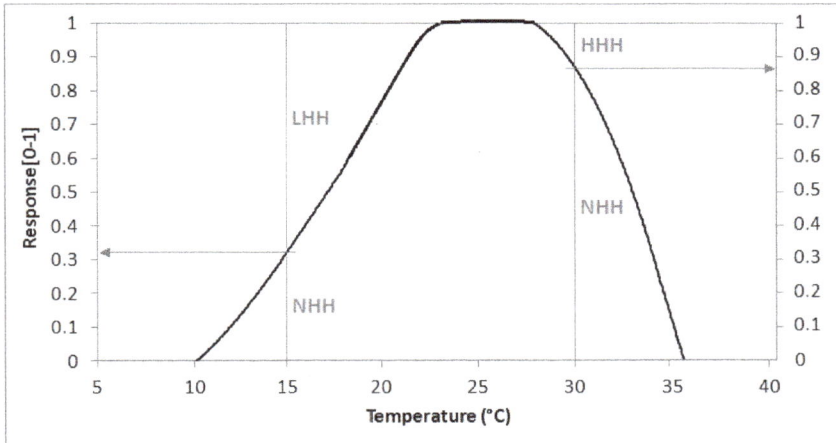

Figure 2. A response curve derived from grapevine which allows translating an hour spent at a given temperature into a fraction of normal heat hours (NHH) (range 0–1) while the complement to 1 represents normal hours which are useless because they are sub-optimal (LHH) or supra-optimal (HHH). For example, an hour spent at 5 °C gives 0 NHH and 1 LHH, an hour spent at 15 °C gives 0.32 NHH and 0.68 LHH and finally an hour spent at 30 °C is 0.87 NHH and 0.13 HHH.

NHH, LHH and HHH values for a given mean hourly temperature (ATH) can be obtained by means of the following simple algorithm (statements are given in Pascal pseudo-language code).

```
if (ATH ≤ LC) or (ATH ≥ UC) then NHH = 0 else
if (ATH ≥ LO) and (ATH ≤ UO) then NHH = 1 else
if (ATH > LC) and (ATH < LO) then
begin
m: = 1/(LO − LC);
q: = 1 − LO/(LO − LC)
NHH = m * ATH + q
end else
if (val1 > UO) and (val1 < UC) then
begin
m: = −1/(UC − UO)
q: = 1 + UO/(UC − UO)
NHH = m * ATH + q
end
if ATH < LO then LHH = 1 − NHH
if ATH > UO then HHH = 1 − NHH
```

This algorithm is based on a theoretical response curve with a trapezoidal shape. The real curve is the result of the whole set of physiological processes ruled by temperature, which means that the availability of experimental curves could be important for enhancing the proposed approach.

GDD or NHH cumulated from the sowing date to a given day can be used to simulate the phenological stage (flowering, fruit-set, ripening and so on), cumulative biomass. or some derived indexes such as the LAI [35]. The scatterplot in Figure 3 shows the relation between values of NHH and GDD without truncation calculated for 202 Italian stations in 2017. The shape of the cloud of points are the consequence of the less physiological approach based on GDD subject to an overestimation growing with the increase of thermal resources. The scatterplot in Figure 4 also from 2017 shows the relation between NHH and Global Solar Radiation for the same stations analyzed in Figure 3. The poor correlation is a result of high altitude stations characterized by high values of radiative resources in coincidence with low or null NHH.

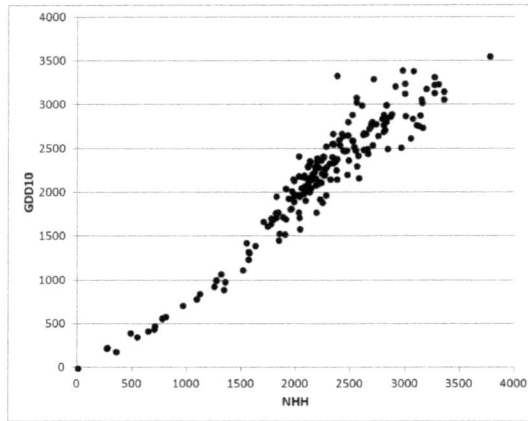

Figure 3. The scatterplot shows the relationship between NHH [h] and growing degree days (GDD) [°C] calculated for a minimum cardinal of 10 °C without truncation. Data are from 2017 for 202 stations in Italy located at elevations between 0 and 3488 m and latitudes between 35.498° N and 46.943° N (our elaborations on NOAA-GSOD data). NHH was calculated with the algorithm based on a response curve with a trapezoidal shape described in the text.

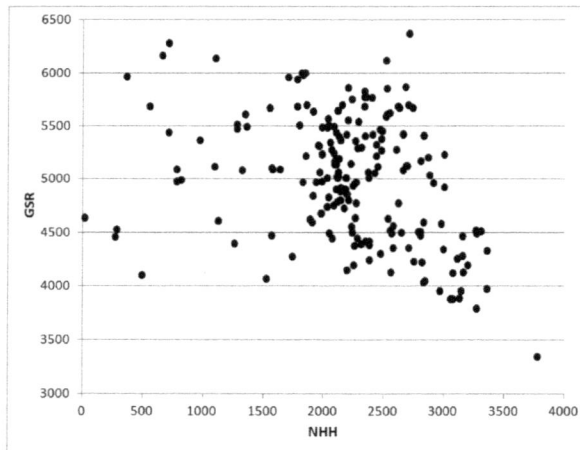

Figure 4. The scatterplot shows the correlation between values of NHH [h] and Global Solar Radiation [MJ m^{-2}]. Data are from 2017 for 202 stations in Italy located at elevations between 0 and 3488 m and latitudes between 35.498° N and 46.943° N (our elaborations on NOAA-GSOD data). NHH was calculated with the algorithm based on a response curve with a trapezoidal shape described in the text.

Temperature is also an important source of information for plants because it indicates the season, allowing optimization of their phenological rhythm in relation to thermal, radiative, and pluviometric features. An example of this is the vernalization process to which several species of medium-high latitudes are subjected. Exposure to low temperatures for a certain period of time, is essential for spring vegetative recovery and the development of vegetative and reproductive organs. Chilling is expressed as chill units (CU), and calculated in the same way as NHH, but by adopting an LC = 0 °C, LO = UO = 7 °C, and UC = 14 °C. CU can be obtained with the same algorithm previously described for NHH calculation. In many species vernalization and photoperiod cooperate to induce flowering in periods suitable for thermal and water resources [36].

Particular attention should be paid to temperatures below the critical threshold CL. By this, frost tolerant species when in full vegetative rest (tissues hardened which means low water content and high concentrations of cellular soluble solids) are able to withstand very low temperatures, such as critical low temperatures of −9/−10 °C for olive (*Olea europaea* L.), −15/−18 °C for grapevine and −20/−22 °C for wheat. It is more difficult to establish the upper critical threshold CH because heat stress from high temperatures depends primarily on canopy temperature rather than on air temperature. So, the interplay between various environmental variables (soil water content, relative humidity, wind speed and global solar radiation) and the stomatal response of plants affects leaf temperatures which can vary substantially for the surrounding air temperature. In general, as stated by De Boeck et al. [37], leaves tend to heat up when stomatal conductance is low (drought conditions) and this effect is strengthened by high solar radiation and high relative humidity, while high wind speeds brings the leaf temperature closer to the air temperature, which can imply either cooling or warming (i.e., abating or reinforcing heat stress) depending on other prevailing conditions. Furthermore, the effect of wind is more relevant with small leaves which have a reduced boundary layer and oscillate easily, enhancing the exchanges of heat with air. For this reason, plants affected by heat stress tend to develop more small leaves than plants growing in environments without stress.

Higher temperatures change root architecture, acting on primary root elongation rate and the overall shape of the root system giving a shallower and broader root distribution and a general increase in lateral root branching as stated by Gray and Brady [38]. The same authors noted that reproductive growth was altered by heat stress for rice which showed an optimum at 33 °C for vegetative growth, while grain formation and yield were negatively affected by temperatures above 25 °C. Moreover temperatures above 33 °C reduced viability of pollen which reached zero for temperatures of 40 °C with a similar phenomenon for sorghum (optimum at 26–34 °C for vegetative growth and at 25–28 °C for reproductive growth) and for Arabidopsis (*Arabidopsis thaliana*), where the abortion of the whole inflorescence was observed at a temperature of 36 °C. Gray and Brady [39] emphasized the cellular response to temperature stress which includes altered organization of organelles, cytoskeleton, and membrane structure. To maintain membrane stability and normal cellular functions in the presence of heat stress, plants synthesize heat shock proteins (HSPs), molecular chaperones that prevent protein misfolding or aggregation, as well as other co-chaperones, hormones, and other protective molecules. Expression of HSPs is induced by heat-stress transcription factors (HSFs) that bind to heat shock elements in the promoters of HSPs. There are many steps of regulation allowing dynamic control of the heat stress response, as the HSFs themselves can be post-transcriptionally modified. In addition to the constitutive role that HSPs play in heat stress responses across cell types, these proteins can acquire specialized functions that regulate developmental responses of organs to environmental stress.

3.4. Agronomic Approach to Thermal Resources and Limitations

The choice of agronomic solutions should be based on a sufficiently detailed territorial agroclimatic analysis, and finalized to quantify thermal resources and limitations for different environments. In such an analysis, it is appropriate to consider the traditional agronomic choices typical of the selected territory because in many cases they are the consequence of a secular adaptation that has discarded fewer effective solutions. In this sense, for example, the Italian alpine territories have a traditional use for forestry and pasture for the slopes exposed to the north, while those exposed to the south are primarily used for fruiting or viticulture which are concentrated in the thermal belt, as in great valleys such as the Valtellina, Valle d'Aosta, and Adige River Valleys, which are located between 250 and 750 m above sea level (asl). Similarly, the Apennine territories have a typical summer crops of maize, sunflower (*Helianthus annuus*), and tobacco (*Nicotiana tabacum*) in the valley bottoms, most exposed to winter and spring frost, while grapevines are located on the lower slopes, at intermediate risk of frost, and finally olive, most sensitive to frost, is traditionally located in the thermal belt which in the Apennines is often between 150 and 350 m asl.

Agroclimatic analysis must be founded on meteorological data of at least 20–30 years close to the present time to provide information representative of the current climate. In the light of resources and limitations (not only the average but also the absolute minimum and maximum and the probability of yearly values lower or higher than specific thresholds), agronomic practices useful for optimizing resources and minimizing limitations can be employed.

Almost the entire set of agronomic practices (soil cultivation, pruning, weeding, fertilization, irrigation, and so on) influence temperatures by interfering with the terms of the energy balance of the field. Thus, only the most important decisions for optimizing thermal resources and minimizing thermal limitations in a given field/territory will be reported. They can be subdivided into strategic decisions (that involve farm activity for many years) and tactical actions (that involve the farm activity only for the current year). Obviously, all these decisions should be founded on both technical and economic evaluation.

The first fundamental decision is the right choice of species and varieties best suited to the planting area, and this should be done after evaluating the length of the growing season and the mean and extreme dates of the beginning of the main phenological stages (e.g., dates of bud-break and ripening must be compatible with the dates of the last killing frost), and the compatibility with the temperature resources (chill units for vernalization, NHH or GDD needed to close the crop cycle), and limitations (LTL, HTL).

The prevention of cold damage should be founded on the uniformity of field morphology to prevent cold air storage which may occur due to the presence of depressions and irregularities in the field slope. It is also important to have suitable openings in barriers (drywalls, compact rows of trees or bushes, etc.) that limit the drainage of cold air enhancing the risk of frost. The adoption of active frost protection systems (low volume sprinkler irrigation over or under-canopy, stoves, candles, etc.) is a relevant decision that should be taken with a careful evaluation of the installation and handling costs and staffing requirements related to system operability.

For poorly drained fields, where a relevant limitation is water excess during winter that slows the spring warming of soils, the hydraulic arrangement of the terrain with ditches or subsurface drains is of paramount importance.

The choice of suitable training systems is important, both for the prevention of cold and heat damage. In fact, the height of the canopies of grapevines and fruit trees is directly related to the risk of frost damage because the coldest layers are closest to the ground. Furthermore, fruit damage from heat should be avoided by adopting training systems that protect fruit with a layer of leaves mitigating thermal extremes.

The risk of high temperatures should be evaluated during planning of harvest activities to prevent loss of quantity and quality of products. In this regard, consideration of options may include harvest machinery, work organization and machinery utilization. For example, harvesting grapes at night, use of refrigerated bins for transport of harvested clusters, and a quick crushing soon after harvesting are solutions sometimes adopted for hot climate viticulture to prevent unwanted fermentation.

Among the tactical decisions, making a rational choice regarding the sowing or transplanting periods for annual herbaceous species and of the planting times for perennial crops (trees, shrubs and herbaceous species) is also important. This decision should be founded on weather forecast information derived from good quality climatological data.

It should not be forgotten that, in conditions of water scarcity, plants close their stomata and the cooling of the foliar tissue are entrusted to the air. In this case, it is useful to plant the crops in rows parallel to the dominant winds to favor the penetration of air into the canopies.

The appropriate use of irrigation is also important in order to limit the negative effects of high temperatures on field crops. The mitigating effect of irrigation on maximum temperatures at a territorial scale was described by Lobell and Bonfils [39] who, working on a long time series of air temperatures for California (1934–2002), highlighted the strong mitigating effect of irrigation that was quantified in a 5 °C reduction in maximum temperature of the entire surface of the State had it been irrigated.

A melon (*Cucumis melo* L.) crop is transplanted early in spring when air and soil temperatures reach 14 °C in order to harvest the fruit early to get the highest price on the market. Lower temperatures can induce chilling injury and delay the recovery and the production of the plants [40]. The agronomic strategy to reduce low temperature damage is to use mulches or rowcovers [41]. Mulches with black or white plastic films can reduce thermal excursion. A black film can increase the soil temperature and increase G&D shortening the production cycle. Rowcovers can increase the soil and air temperature around the transplanted plants.

In summer, high temperatures can accelerate G&D of the plants too much, inducing a fast ripening of fruits with a shortened growing cycle. Such fast growth usually reduces the quality of fruis since the photosynthetic period is short and the sugar loading in fruit is reduced, resulting in low quality fruit. The best strategy is to avoid this problem is to identify the best genotype, cultivar or variety, since there are differences in sugar accumulation during ripening related to sucrose phosphate synthase and acid invertase activities [42]. In the open field, another agronomic strategy to reduce the negative effect of high temperature is to improve thermoregulation by keeping the soil water content high, if there is adequate water availability.

Thermal stress conditions can be detected by measurements carried out with thermal infrared cameras managed by hand or installed on drones or satellites. Symptoms of thermal stress can also occur in the presence of pests and diseases that alter stomatal functionality which is essential for plant thermal regulation.

Table 1. Cardinal temperatures for some crops with the sources of information used.

Common Name	Scientific Name	Minimum Cardinal	Optimal Range	Maximum Cardinal	Reference
Alfalfa	*Medicago sativa* L.	8	24–26	36	[43]
Asparagus	*Asparagus officinalis* L.	4	18–22	28	[44]
Banana	*Musa* seppe. L.	12	25–30	40	[43]
Barley	*Hordeum vulgare* L.	2	18–28	34	[45]
Bean	*Phaseolus vulgaris* L.	10	24–30	36	[45]
Carrot	*Daucus carota* L.	3	16–22	28	[44]
Cotton	*Gossypium hirsutum* L.	14	25–30	38	[33]
Durum wheat	*Triticum durum* L.	2	18–26	32	[44]
Flax	*Linum usitastissimum* L.	2	18–24	30	[44]
Grapevine	*Vitis vinifera* L.	7–10	22–28	36	[26]
Lemon	*Citrus limon* L.	13	23–30	35	[43]
Maize	*Zea mays* L.	8	22–30	35	[44]
Melon	*Cucumis melo* L.	15	25–35	40	[44]
Oat	*Avena sativa* L.	2	18–26	32	[44]
Okra	*Abelmoschus esculentus* L.	16	25–35	40	[44]
Olive	*Olea europaea* L.	10	22–28	38	[43]
Onion	*Allium cepa* L.	2	20–28	34	[43]
Sweet potato	*Hipomea batata* L.	15	25–33	38	[44]
Peanuts	*Arachis ipogea* L.	11	23–30	40	[33]
Pineapple	*Ananas comosus* L.	15	22–30	35	[43]
Pea	*Pisum sativum* L.	4	15–20	30	[43]
Tomato	*Solanum lycopersicum* L.	12	22–26	35	[33]
Potato	*Solanum tubrerosum* L.	4	14–23	33	[46]
Rapeseed	*Brassica napus oleifera* L.	5	15–20	30	[47]
Rice	*Oryza sativa* L.	12	25–32	38	[33]
Rye	*Secale cereale* L.	2	20–26	31	[44]
Safflower	*Carthamus tinctorius* L.	10	18–28	35	[43]
Soft wheat	*Triticum aestivum* L.	2	18–26	32	[45]
Sorghum	*Sorghum bicolor* L.	12	24–30	36	[43]
Soybean	*Glycine max* L.	10	20–28	34	[43]
Strawberry	*Fragaria X ananassa* L:	4	15–20	28	[44]
Sugar Beet	*Beta vulgaris* L.	2	18–24	30	[47]
Sugarcane	*Saccharum officinarum* L.	15	22–30	35	[43]
Sunflower	*Helianthus annuus* L.	7	18–25	35	[43]
Tobacco	*Nicotiana tabacum* L.	15	22–30	38	[43]
Watermelon	*Citrullus lanatus* L.	12	22–30	35	[43]

4. Relative Humidity and Effects on Crops

Plants are insensitive to the absolute atmospheric water content expressed for example as mixing ratio or absolute humidity [4,48], while they are quite sensitive to relative humidity (RH) which at a given temperature is the water content of the atmosphere expressed as a percentage of the saturated water content, which is a constant at a given temperature. RH is an important environmental variable for crop productivity, because it regulates the transpiration rate at the leaf level and can influence the water balance in crops. A high RH limits transpiration and reduces growth and nutrient assimilation. A low RH increases water flux through plants and increases transpiration with severe problems in species with a reduced ability to regulate stomatal aperture.

4.1. Space and Time Variability of Relative Humidity

The RH in a plant canopy is the result of a balance between humidity received from soil evaporation (which is enhanced when the soil surface is well-watered by rainfall, irrigation or by lifting from water tables) and plant transpiration (which in the absence of soil water limitation is a function only of the atmosphere and canopy features). Wind is crucial for removing water from canopy layers, including breezes and local winds induced by lack of thermal homogeneity among land surfaces, sea, lakes, forests, swamps, cropping areas, etc. During the day lack of homogeneity is induced by differential solar heating of surfaces with different characteristics that trigger the establishment of stationary convective cells in the planetary boundary layer (first 1000–1500 m asl), while at night it is induced by differential radiative cooling with production of cold air pools that drain along the relief. Another important determinant of RH is foehn, a katabatic wind typical of areas located downwind of mountain chains [49]. Foehns cause substantial drops in RH producing stress conditions for crops mainly due to the substantial increase of water demand of the atmosphere. Also relevant to RH are other macroscale and circulation patterns typical of different tropical and mid latitude contexts; these are beyond the scope of this review but can be appreciated in meteorological treatises [50]. In the specific case of dynamic anticyclone weather patterns, which for Italy accounts for 50–60% of the total days of the year, the daily pattern of RH shows a regular behavior with a nighttime minimum in coincidence with low values of air temperature and a daily maximum in coincidence with high values of air temperature. Nevertheless, while AT shows a behavior more similar to a sinusoid with minimum at sunrise and maximum 3–4 h before sunset [51,52], RH shows two abrupt changes with an abrupt decrease in coincidence with the beginning of convective vertical exchanges between a canopy layer and upper layers triggered by solar heating in the morning and an abrupt increase with the fade out of vertical exchanges (disruption of convective cells) in coincidence with the evening decrease of solar radiation before sunset.

4.2. Effects of Low/High Relative Humidity on Crops

Low RH increases evapotranspiration, enhancing water needs of rainfed and irrigated crops and, consequently, the risk of water stress conditions due to the lack of easily accessible water in the area explored by roots. On the other hand, high values of RH reduce the quantity of transpired water [51], which reduces soil water stress. Nevertheless, a high RH in the presence of high values of solar radiation can create problems of thermal excess [53], because plants control the temperature of their tissues by transpiring water, which changes its state from liquid to vapor and removes 2450 J g^{-1} from plant tissues in the evaporated water.

It must also be considered that many organic substances like starch are hygroscopic, which means that grains in presence of high relative humidity are re-hydrated. Multiple cycles of drying/re-hydration enhance the cracking phenomenon in rice with a significant reduction in commercial value of the product [54].

RH plays an important role in plant nutritional status because plants exposed to high RH show two contrasting effects on growth. On the one hand, plants might show increased growth due to higher stomatal opening, leading to increased uptake of CO_2. On the other hand they might show reduced growth due to a reduced transpiration volume leading to a lower translocation of nutrients [55].

High values of RH are favorable to parasitic fungi with a direct effect on their growth and development and an indirect one on the family of Peronosporaceae that needs the presence of condensed water on plant organs (leaves and shoots) for the activity of zoospores which is an essential part of their cycle. Consequently, many agronomic practices have as main or secondary objectives the avoidance of a long persistence of excessively low or high values of RH inside the canopy. Among these practices are:

1. The choice of areas not excessively humid, avoiding for example valley bottoms or basins with low air movement;
2. The adoption of suitable training systems for grapevine and other fruit crops;
3. The adoption of suitable plant distances along and between rows;
4. The execution of suitable tillage practices that enhance air circulation inside the canopy (e.g., weed management, cutting of grass in orchards and vineyards, leaf removal, winter or green pruning, etc.).

Dew and hoarfrost are the result of water condensation in the presence of a saturated atmosphere on objects (plants, rocks, and others) which have a temperatures below the dew point or the 0 °C threshold. The contribution of dew or hoarfrost to satisfaction of the water needs of crops is generally low because the dew accumulating in a single night at mid latitudes is often below 0.2 mm [56,57] while the mean daily water need for a reference crop in summer is about 4–6 mm. Dew contribution may be important in areas located close to large sources of humidity like seas or lakes and with high daily thermal ranges where water from dew can exceed 0.5–1 mm which means 0.5–1 L m^{-2}.

5. Future Prospective and Conclusions

Understanding the sources of abiotic stresses, how plants respond to them for improving tolerance, and the use of specific agronomy strategies for stress alleviation is essential. With reference to the scheme of the production model in Figure 2 of Mariani and Ferrante [1], the present analysis of resources and limitations related to light radiation, temperature, and RH has been performed. This analysis follows the analysis of abiotic factors carried out in Mariani and Ferrante [1].

It is important to consider that crop management needs an overall view of the atmospheric variables and the meteorological events that occur in a given year. For example, in the case of the temperate rice cropping areas (about the 35% of the total rice world area), it is well known that:

1. Low values of global solar radiation and suboptimal/supra-optimal temperatures reduce photosynthesis, while low daily thermal ranges reduced the translocation of biomass from leaves to storage organs:
2. Water for paddy rice submersion can be insufficient in years with low levels of water resources;
3. Long rainy periods are serious obstacles to the rational management of weeds, pests, and pathogens, preventing timely spraying of pesticides, fungicides, and herbicides;
4. Low temperatures during floral differentiation enhance rice male sterility with drastic reduction of yields. The critical threshold for rice sterility is 12 °C and farmers generally counteract this problem by increasing the water level in rice ponds;
5. Long periods of high temperatures (mean daily values above the 26 °C threshold) enhance rice cracking and rice chalking, phenomena that reduce significantly the quality of the final product and that are also strictly related to varieties and nutrient management;
6. Meteorological variables like temperature, relative humidity, and rainfall have a strong influence on rice pests, weeds, and pathogens.

Diseases impose a series of strategic and tactical choices suitable for preventing and controlling them, ranging from the choice of varieties and sowing periods to the strategies of water, pest, and weed management. These strategies have evolved over the centuries, and have been refined based on the available technologies and the experience acquired by the farmers, in response to the constant interaction with the climate which characterizes the area under examination. For the main European rice cropping area, located in the north of Italy (where rice has been cultivated since the 15th century), an historical document of great value for appreciating this aspect is provided by the correspondence between Camillo Benso, Count of Cavour, and his partner Giacinto Corio [58,59] who managed the three large farms of Leri, Torrone, Montarucco with a total area of more than 400 hectares. From these letters emerge the profound agronomic culture that characterizes not only Giacinto Corio, but also Camillo Benso who was then prime minister of the kingdom of Piedmont and the main father of the unification of Italy. Today, this agronomic culture derived from constant interaction of the farmer with atmospheric phenomena is a key element of global food security, and the task of science is to make it more rational and based on qualitative data and interpretative and predictive models. The mitigation of stressful conditions can be achieved using appropriate agronomic management practices that have to be chosen harmonizing all production factors in a specific area with well known pedo-climatic conditions.

In this two part review, we have analyzed some relevant abiotic stress factors (drought, hypoxia, and lodging in the first part [1], and temperature, light intensity, and relative humidity in this second part). We have described the stress factors, and the agronomic strategies for mitigation and adaptation to them. From this we can derive some suggestions to guide future approaches to such abiotic stresses.

Firstly, it is necessary to improve our knowledge of the main stress factors in each production area, including their average incidence, and their variability in space and time. These analyses are now feasible by the availability of monitoring tools for agroclimatic conditions, soils, and crops. For the future a more systematic approach to these analyses is needed. It is of paramount importance that a more general adoption of techniques for timely monitoring of crop performance and pedoclimatic conditions (e.g., agrometeorological stations and remote sensing techniques based on satellites and drones) that allow defining the mesoscale and microscale variability of the various stress factors be established. In addition, the availability of time series of meteorological, phytopathological and productive data is also essential. These series at the regional level should be available to farmers and technicians by authorities. Farmers should take care of collection and archiving of time series within their farm.

Another essential factor to promote is the knowledge of the ability of different species and varieties to adapt to stress factors to be able to counteract the stress with a suitable varietal choice and crop succession. Genetic improvement of crops must be accompanied by the adoption of suitable and innovative cultivation techniques (mechanization, training systems for orchards and vineyards, hydraulic-agrarian arrangements, tillage, irrigation, fertilization, weeding, pest control, etc.).

The choice of the optimal times for the execution of the different field activities (times for soil cultivation, seeding periods, and harvesting) is another crucial factor to limit the negative effects of stress.

The adoption of precision farming techniques is useful for dominating the microscale variability of the different stress factors. Moreover, the adoption of conservative agriculture practices allows maintaining the levels of the resources within the limits of acceptability for different crops (e.g., dry farming).

Use of computerized decision support systems (DSS), and extension activities, are essential for engaging farmers in the process of innovation in genetics and crop production techniques. This aspect is more important considering that at the world level agriculture is carried out by 590 million farms with characteristics that are extremely diversified by size, number of workers, levels of mechanization, and openness to the market.

It is also important to maintain a central role for agronomy seen as the science of cultivation. This discipline should maintain its own centrality as a science that guides the producer to optimize their results in terms of quantity and quality of the products obtained.

A further element worthy of comment is the interaction between agronomic techniques and genetic innovation aimed at the improvement of the resistance to abiotic stresses. A complementary relationship should generally be desirable because every new variety needs to be embedded in an agronomic context favorable to the full expression of its potential resistance to abiotic stresses. Furthermore, agronomic techniques can be a realistic way to mitigate the negative effects of abiotic stresses while strategic actions of genetic improvement are carried out and the new varieties are available to farmers. In conclusion, we hope that this analysis, obviously somewhat brief given the vastness of the topic, can help those who for different purposes are concerned with the determinants of agricultural production.

Author Contributions: Temperature and relative humidity, L.M.; solar Radiation A.F. and L.M.; writing A.F. and L.M.

Funding: This research received no external funding.

Conflicts of Interest: The authors declare no conflict of interest.

References

1. Mariani, L.; Ferrante, A. Agronomic Management for Enhancing Plant Tolerance to Abiotic Stresses—Drought, Salinity, Hypoxia, and Lodging. *Horticulturae* **2017**, *3*, 52. [CrossRef]
2. Boyer, J.S. Plant productivity and environment. *Science* **1982**, *218*, 443–448. [CrossRef] [PubMed]
3. Farooq, M.; Wahid, A.; Kobayashi, N.; Fujita, D.; Basra, S.M.A. Plant drought stress: Effects, mechanisms and management. *Agron. Sustain. Dev.* **2009**, *29*, 185–212. [CrossRef]
4. Stull, R. *Practical Meteorology: An Algebra-Based Survey of Atmospheric Science*; Version 1.02b; University of British Columbia: Vancouver, BC, Canada, 2017; p. 940, ISBN 978-0-88865-283-6. Available online: https://www.eoas.ubc.ca/books/Practical_Meteorology/ (accessed on 15 July 2018).
5. Long, S.P.; Bernacchi, C.J. Gas exchange measurements, what can they tell us about the underlying limitations to photosynthesis? Procedures and sources of error. *J. Exp. Bot.* **2003**, *54*, 2393–2401. [CrossRef] [PubMed]
6. Hagemann, M.; Weber, A.; Eisenhut, M. Photorespiration: Origins and metabolic integration in interacting compartments. *J. Exp. Bot.* **2016**, *67*, 2915–2918. [CrossRef]
7. Osborne, C.P.; Beerling, D.J. Nature's green revolution: The remarkable evolutionary rise of C4 plants. *Philos. Trans. R. Soc. Lond. B Biol. Sci.* **2006**, *361*, 173–194. [CrossRef] [PubMed]
8. Osborne, C.P.; Freckleton, R.P. Ecological selection pressures for C4 photosynthesis in the grasses. *Proc. R. Soc. Lond. B Biol. Sci.* **2009**, *276*, 1753–1760. [CrossRef] [PubMed]
9. Greaves, G.E.; Wang, Y.M. The effect of water stress on radiation interception, radiation use efficiency and water use efficiency of maize in a tropical climate. *Turk. J. Field Crops* **2017**, *22*, 114–125. [CrossRef]
10. Kadir, S.; Von Weihe, M. Photochemical efficiency and recovery of photosystem ii in grapes after exposure to sudden and gradual heat stress. *J. Am. Soc. Hortic. Sci.* **2007**, *132*, 764–769.
11. Parent, B.; Tardieu, F. Can current crop models be used in the phenotyping era for predicting the genetic variability of yield of plants subjected to drought or high temperature? *J. Exp. Bot.* **2014**, *65*, 6179–6189. [CrossRef] [PubMed]
12. Bouman, B.A.M.; van Keulen, H.; van Laar, H.H.; Rabbinge, R. The 'School of de Wit' crop growth simulation models: A pedigree and historical overview. *Agric. Syst.* **1996**, *52*, 171–198. [CrossRef]
13. Avnish Kumar, B. *Crop Growth Simulation Modeling, in Modelling and Simulation of Diffusive Processes*; Springer: Berlin, Germany, 2014; pp. 315–332.
14. Wagenmakers, P.S. *Light Relations in Orchard Systems*; Wageningen University: Wageningen, The Netherlands, 1995; ISBN 90-5485-340-9. Available online: http://edepot.wur.nl/205150 (accessed on 15 July 2018).
15. Wertheim, S.J. Productivity and fruit quality of apple in single-row and fullfield plantings. *Sci. Hortic.* **1985**, *26*, 191–208. [CrossRef]
16. Hirose, T. Development of the Monsi-Saeki Theory on Canopy Structure and Function. *Ann. Bot.* **2005**, *95*, 483–494. [CrossRef] [PubMed]
17. Williams, L.E.; Ayars, J.E. Grapevine Water Use and the Crop Coefficient are Linear Functions of the Shaded Area Measured beneath the Canopy. *Agric. For. Meteorol.* **2005**, *132*, 201–211. [CrossRef]

18. Woods, D.P.; Ream, T.S.; Minevich, G.; Hobert, O.; Amasino, R.M. Phytochrome C Is an Essential Light Receptor for Photoperiodic Flowering in the Temperate Grass Brachypodium distachyon. *Genetics* **2014**, *198*, 397–408. [CrossRef] [PubMed]

19. Quail, P.H. Phytochrome photosensory signaling networks. *Nat. Rev. Mol. Cell Biol.* **2002**, *3*, 85–93. [CrossRef] [PubMed]

20. Pham, V.N.; Kathare, P.K.; Huq, E. Phytochromes and phytochrome interacting factors. *Plant Physiol.* **2018**, *176*, 1025–1038. [CrossRef] [PubMed]

21. Foyer, C.H.; Lelandais, M.; Kunert, K.J. Photooxidative stress in plants. *Physiol. Plant.* **1994**, *92*, 696–717. [CrossRef]

22. Trivellini, A.; Cocetta, G.; Francini, A.; Ferrante, A. Reactive oxygen species production and detoxification during leaf senescence. In *Reactive Oxygen Species and Antioxidant Systems in Plants: Role and Regulation under Abiotic Stress*; Khan, M.I.R., Khan, N., Eds.; Springer Nature Singapore Ltd.: Singapore, 2017; pp. 115–128.

23. Cola, G.; Failla, O.; Mariani, L. BerryTone, a simulation model for the daily course of grape berry temperature. *Agric. For. Meteorol.* **2009**, *149*, 1215–1228. [CrossRef]

24. Iglesias, I.; Alegre, S. The effect of anti-hail nets on fruit protection, radiation, temperature, quality and profitability of 'Mondial Gala'apples. *J. Appl. Hortic.* **2006**, *8*, 91–100.

25. Melgarejo, P.; Martınez, J.J.; Hernández, F.; Martınez-Font, R.; Barrows, P.; Erez, A. Kaolin treatment to reduce pomegranate sunburn. *Sci. Hortic.* **2004**, *100*, 349–353. [CrossRef]

26. Mariani, L.; Failla, O. *Clima e Viticoltura, Capitolo 2 del Testo Progressi in Viticoltura, a Cura di M. Boselli*; EDISES Universitaria: Napoli, Italy, 2016; pp. 19–38.

27. Annunziata, M.G.; Apelt, F.; Carillo, P.; Krause, U.; Feil, R.; Mengin, V.; Lunn, J.E. Getting back to nature: A reality check for experiments in controlled environments. *J. Exp. Bot.* **2017**, *68*, 4463–4477. [CrossRef] [PubMed]

28. Zhao, X.Y.; Yu, X.H.; Liu, X.M.; Lin, C.T. Light regulation of gibberellins metabolism in seedling development. *J. Integr. Plant Biol.* **2007**, *49*, 21–27. [CrossRef]

29. Grebe, M. Out of the shade and into the light. *Nat. Cell Biol.* **2011**, *13*, 347. [CrossRef] [PubMed]

30. Halliday, K.J.; Fankhauser, C. Phytochrome-hormonal signalling networks. *New Phytol.* **2003**, *157*, 449–463. [CrossRef]

31. Woodrow, P.; Ciarmiello, L.F.; Annunziata, M.G.; Pacifico, S.; Iannuzzi, F.; Mirto, A.; Carillo, P. Durum wheat seedling responses to simultaneous high light and salinity involve a fine reconfiguration of amino acids and carbohydrate metabolism. *Physiol. Plant.* **2017**, *159*, 290–312. [CrossRef] [PubMed]

32. Larcher, W. *Physiological Plant Ecology*, 3rd ed.; Springer: Berlin, Germany, 1995; p. 506.

33. Luo, Q. Temperature thresholds and crop production: A review. *Clim. Chang.* **2011**, *109*, 583–598. [CrossRef]

34. Mariani, L. *Agronomia*; CUSL: Milano, Italian, 2014; p. 344. (In Italian)

35. Mariani, L. Carbon plants nutrition and global food security. *Eur. Phys. J. Plus* **2017**, *132*, 69. [CrossRef]

36. Körner, C.; Basler, D. Phenology under Global Warming. *Science* **2012**, *327*, 1461–1462. [CrossRef] [PubMed]

37. De Boeck, H.J.; Van De Velde, H.; De Groote, T.; Nijs, I. Ideas and perspectives: Heat stress: More than hot air. *Biogeosciences* **2016**, *13*, 5821–5825. [CrossRef]

38. Gray, S.B.; Brady, S.M. Plant developmental responses to climate change. *Dev. Biol.* **2016**, *419*, 64–77. [CrossRef] [PubMed]

39. Lobell, D.B.; Bonfils, C. The Effect of Irrigation on Regional Temperatures: A Spatial and Temporal Analysis of Trends in California, 1934–2002. *J. Clim.* **2008**, *21*, 2063–2071. [CrossRef]

40. Korkmaz, A.; Dufault, R.J. Developmental consequences of cold temperature stress at transplanting on seedling and field growth and yield. II. Muskmelon. *J. Am. Soc. Hortic. Sci.* **2001**, *126*, 410–413.

41. Jenni, S.; Stewart, K.A.; Cloutier, D.C.; Bourgeois, G. Chilling injury and yield of muskmelon grown with plastic mulches, rowcovers, and thermal water tubes. *HortScience* **1998**, *33*, 215–221.

42. Hubbard, N.L.; Huber, S.C.; Pharr, D.M. Sucrose phosphate synthase and acid invertase as determinants of sucrose concentration in developing muskmelon (*Cucumis melo* L.) fruits. *Plant Physiol.* **1989**, *91*, 1527–1534. [CrossRef] [PubMed]

43. Schultink, G.; Amaral, N.; Mokma, D. *Users Guide to the CRIES Agro-Economic Information System Yield Model*; Michigan State University: East Lansing, MI, USA, 1987; p. 125.

44. USDA (United States Department of Agriculture). *Technical Bulletin*; USDA: Washington, DC, USA, 1977; pp. 1516–1525.

45. Weikay, Y.; Hunt, L.A. An Equation for Modelling the Temperature Response of Plants using only the Cardinal Temperatures. *Ann. Bot.* **1999**, *84*, 607–614.

46. Rykaczewska, K. The Impact of High Temperature during Growing Season on Potato Cultivars with different response to Environmental Stresses. *Am. J. Plant Sci.* **2013**, *4*, 2386–2393. [CrossRef]

47. North Dakota Agricultural Weather Network. *Sunflower Development and Growing Degree Days (GDD)*; Crop and Pest Report 2016; North Dakota State University: Fargo, ND, USA; Available online: https://ndawn.ndsu.nodak.edu/help-sunflower-growing-degree-days.html (accessed on 15 July 2018).

48. Mc Intosh, D.H.; Thom, A.S. *Essentials of Meteorology*; Wikeham Publications: London, UK, 1972; p. 239.

49. Ambrosetti, P.; Mariani, L.; Scioli, P. Climatology of north foehn in Canton Ticino and Western Lombardy. *Riv. Ital. Agrometeorol.* **2005**, *2*, 24–30.

50. Barry, R.G.; Chorley, R.J. *Atmosphere, Weather and Climate*; Routledge: London, UK, 2009; p. 499.

51. Allen, R.G.; Pereira, L.S.; Raes, D.; Smith, M. *Crop Evapotranspiration-Guidelines for Computing Crop Water Requirements*; FAO Irrigation and Drainage Paper 56; Food and Agriculture Organization of the United Nations: Rome, Italy; Available online: http://www.fao.org/docrep/X0490E/X0490E00.htm (accessed on 15 July 2018).

52. Parton, W.J.; Logan, J.A. A model for diurnal variation in soil and air temperature. *Agric. Meteorol.* **1981**, *23*, 205–216. [CrossRef]

53. Hatfield, J.L.; Prueger, J.H. Temperature extremes: Effect on plant growth and development. *Weather Clim. Extremes* **2015**, *10*, 4–10. [CrossRef]

54. Lu, R.; Siebenmorgen, T.J. Modeling rice field moisture content during the harvest season-part 1-model development. *Trans. ASAE* **1994**, *37*, 545–551. [CrossRef]

55. Roriz, M.; Carvalho, S.M.P.; Vasconcelos, M.W. High relative air humidity influences mineral accumulation and growth in iron deficient soybean plants. *Front. Plant Sci.* **2014**, *5*, 726. [CrossRef] [PubMed]

56. Beysens, D.; Muselli, M.; Nikolayev, V.; Narhe, R.; Milimouk, I. Measurement and modelling of dew in island, coastal and alpine areas. *Atmos. Res.* **2005**, *73*, 1–22. [CrossRef]

57. Kabela, E.D.; Hornbuckle, B.K.; Cosh, M.H.; Anderson, M.C.; Gleason, M.L. Dew frequency, duration, amount, and distribution in corn and soybean during SMEX05. *Agric. For. Meteorol.* **2009**, *149*, 11–24. [CrossRef]

58. Fondazione Camillo Cavour. *Lettere di Giacinto Corio a Camillo Cavour (1843–1855)*; Fondazione Camillo Cavour: Santena, Italy, 1980; p. 474. (In Italian)

59. Visconti, E. *Cavour Agricoltore, Lettere Inedite di Camillo Cavour a Giacinto Corio Precedute da un Saggio di Ezio Visconti, G*; Barbera editore: Firenze, Italy, 1913; p. 390.

horticulturae

MDPI

Review

Agronomic Management for Enhancing Plant Tolerance to Abiotic Stresses—Drought, Salinity, Hypoxia, and Lodging

Luigi Mariani [1,2] **and Antonio Ferrante** [2,*]

[1] Lombardy Museum of Agricultural History, via Celoria 2, 20133 Milan, Italy; luigi.mariani@unimi.it
[2] Department of Agricultural and Environmental Sciences, Università degli Studi di Milano, via Celoria 2, 20133 Milan, Italy
*   Correspondence: antonio.ferrante@unimi.it

Academic Editors: Alessandra Francini and Luca Sebastiani
Received: 26 September 2017; Accepted: 21 November 2017; Published: 1 December 2017

Abstract: Abiotic stresses are currently responsible for significant losses in quantity and reduction in quality of global crop productions. In consequence, resilience against such stresses is one of the key aims of farmers and is attained by adopting both suitable genotypes and management practices. This latter aspect was reviewed from an agronomic point of view, taking into account stresses due to drought, water excess, salinity, and lodging. For example, drought tolerance may be enhanced by using lower plant density, anticipating the sowing or transplant as much as possible, using grafting with tolerant rootstocks, and optimizing the control of weeds. Water excess or hypoxic conditions during winter and spring can be treated with nitrate fertilizers, which increase survival rate. Salinity stress of sensitive crops may be alleviated by maintaining water content close to the field capacity by frequent and low-volume irrigation. Lodging can be prevented by installing shelterbelts against dominant winds, adopting equilibrated nitrogen fertilization, choosing a suitable plant density, and optimizing the management of pests and biotic diseases harmful to the stability and mechanic resistance of stems and roots.

Keywords: drought; lodging; hypoxia; salinity

1. Introduction

Crop yield and quality are the result of the interaction between a genotype's potential expression and the environment, which is modified by agronomic management in order to meet the objectives of the farmer. Different genotypes have varying yield capabilities depending on their adaptation abilities [1]. Agricultural systems are continuously evolving due to innovation in agronomic tools and the identification of high-performance cultivars coming from traditional or biotechnological genetic improvements [2]. Abiotic stresses such as low water availability, high salinity, high or low temperatures, hypoxia/anoxia, and nutrient deficiency are among the major causes of crop failure. Plants are able to perceive environmental stimuli and adapt to different environments; however, the degree of tolerance and adaptability to abiotic stresses varies among species and varieties. Crops exposed to abiotic stresses respond by activating defense mechanisms. Therefore, crops in an early stage of stress do not show visible symptoms but their physiology can undergo significant changes [3]. The energy used to counteract or cope with abiotic stresses is called "fitness cost" and does not contribute to crop production. Crops have to balance the resource allocation between productivity and defense actions [4].

In this review, agronomic strategies aimed at optimizing the resilience of crops exposed to abiotic stresses are covered. The work has been divided into two parts and the second part will be related to stress due to nutrients, high and low temperatures, and light excess or deficiency. Every agronomic strategy presented and discussed hereafter is sustainable not only socially and environmentally but also economically, because agriculture is an economic activity that cannot be done without adequate remuneration.

2. Role and Impact of Abiotic Limitations to Crop Yield

The general scheme proposed in Figure 1 describes the plant response to an abiotic stress with given features (duration, severity, etc.). The stress acts on a crop with given phenotypic characters to result in the final effects on growth, development, and mortality rate [5]. The basic scheme of Figure 1 is the result of a complex causal chain, which involves plant hormones and acts at the molecular and physiological level. The knowledge of this causal chain is substantially increasing thanks to transcriptomics, metabolomics, proteomics, and other integrated research approaches [3]. The conceptual scheme illustrated in Figure 2, widely adopted in crop yield simulation [6,7], shows crop production as the final result of a dry matter cascade triggered by solar radiation intercepted by plant canopies, which provides energy for the photosynthetic process.

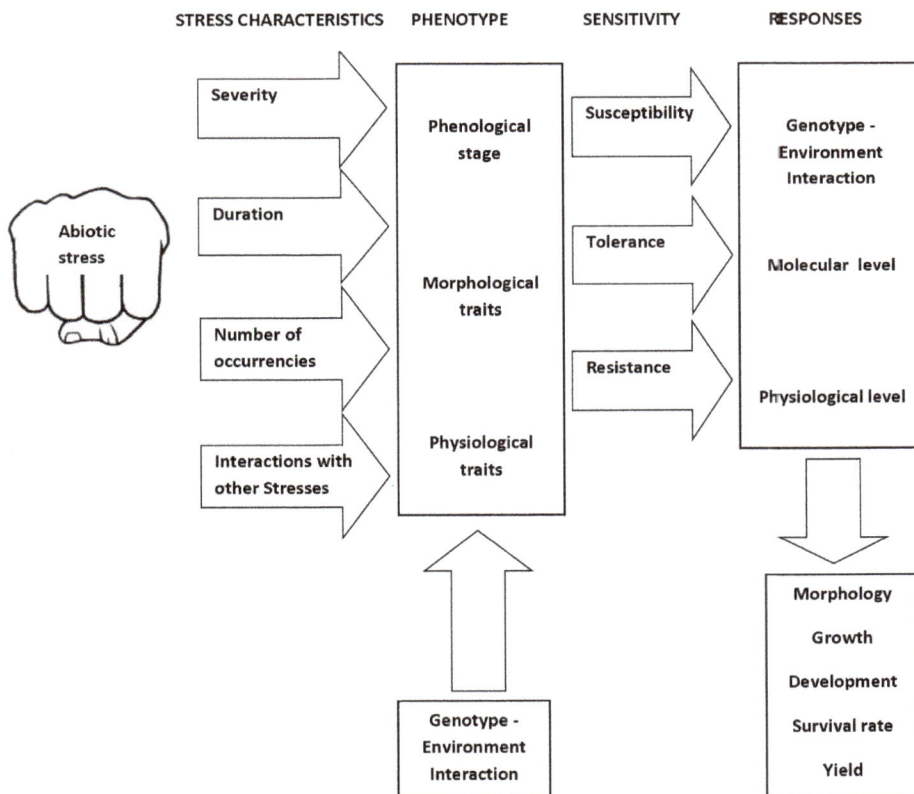

Figure 1. Scheme of factors that determine crop response to abiotic stresses.

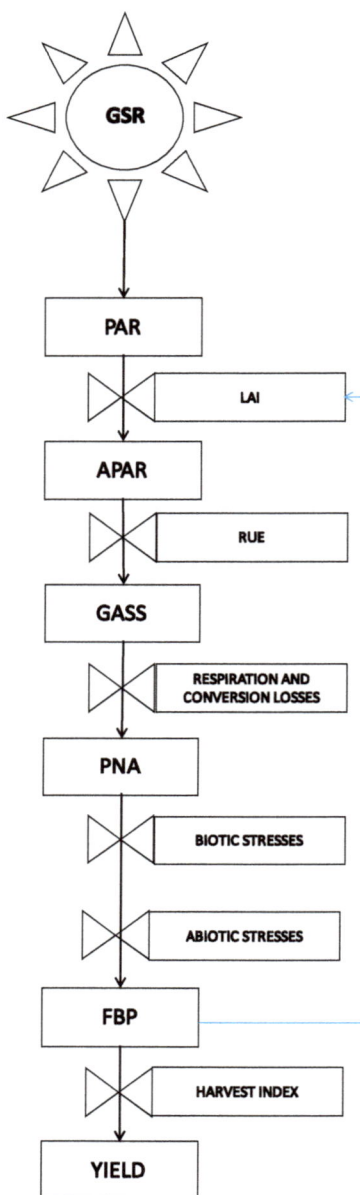

Figure 2. Conceptual scheme of the cascade of energy and organic matter that links global solar radiation and final yield. The role of abiotic stress as a rate variable that rules the conversion from potential net assimilation (PNA) to the final biomass production (FBP) is highlighted, with global solar radiation (GSR), photosynthetically active radiation (PAR), absorbed PAR (APAR), and gross photosynthetic assimilation (GASS) each contributing to the outcome. Rectangular boxes are state variables, and valves are rate variables.

Photosynthesis gives rise to a potential production of dry matter, gradually curtailed by different losses up to final production. Important losses occur by conversion from global solar radiation (GSR) to photosynthetically active radiation (PAR), efficiency of photosynthesis, translocation from photosynthetic to storage organs, maintenance and production respiration, and limitations by biotic (pests, fungi, bacteria, weeds, and so on) and abiotic stresses (temperature, soil water excess or shortage, nutrients, wind, etc.).

An example of a grain maize (class 700 FAO) field crop with Radiation Use Efficiency (RUE) of 4 g of glucose per MJ of Absorbed Photosynthetically Active Rradiation (APAR) and a harvest index (HI) of 0.6 cultivated on a flat plain at 45° North is useful to give an idea of the strength of the effect of different limitations on crop production. This field received 4337.0 MJ·m^{-2} of GSR in the period 1 April–30 September (data for Piacenza San Damiano, Italy—mean of the period 1993–2013); 50% of the GSR was useful for photosynthesis (PAR = 2168.5 MJ) and 80% of the PAR was intercepted by the canopy (APAR = 1734.8 MJ·m^{-2}). By consequence, the Gross Assimilation (GASS) was 1734.8 MJ·m^{-2}·4 g·MJ^{-1}·10,000 m^2·1/1,000,000 t·g^{-1} = 69.4 t·ha^{-1} of glucose.

If we consider a loss of 35% related to maintenance and production respiration and translocation of photosynthetic products from the green organs to the storage tissues [8], there was a total net production without limitations (Potential Net Assimilation, PNA) of 69.4 × (1 − 0.35) = 45.1 t·ha^{-1}, which, multiplied by HI, gives a potential net grain production of 27.1 t·ha^{-1}. This latter value was 48% higher than ordinary production and 33% higher than the maximum production attainable in ideal field conditions (Table 1). These gaps give an idea of the weight of biotic and abiotic stress factors in ordinary field conditions on the Po plain (Italy). In our experience, the weight of biotic factors (mainly effects of the European corn borer *Ostrinia nubilaris* and some fungal diseases) is about 15%, as can be inferred by comparing Italian production trends with those of the USA, where GMO BT maize largely eliminates the effects of biotic stresses from European corn borer and related fungal diseases. As a consequence, the effect of abiotic stresses on maize yield is about 33% (15–48%) in ordinary field conditions and drops to 18% (15–33%) in ideal field conditions (Table 1). Global values of abiotic limitations were simulated by Mariani [8] with a physiological-process-based crop simulation model driven by 1961–1990 monthly climate data from a global FAO dataset and applied to four crops (Wheat, Maize, Rice, and Soybeans (WMRS)) that account for 64% of global caloric consumption by humans. The model simulated only temperature and water limitations.

Table 1. The gap between net potential assimilation and final yield for maize cultivated in the Po valley in Italy [8].

Maize (Class 700 FAO) Yield in Field Conditions on the Po Plain (Italy)	Total Biomass Production (t·ha^{-1})	Harvest Index (%)	Grain Yield (t·ha^{-1})
Potential net assimilation (PNA)	45.1	0.6	27.1
Ordinary farmer objective in field conditions	21.7	0.6	13
Highest yield reachable in field conditions	30	0.6	18

As stated by Mariani [8], (i) the simulation was carried out on a global map with a pixel of 0.50 × 0.50 degrees in geographic coordinates (about 60 × 60 km at the equator), (ii) thermal and water limitations at the different latitudes were estimated only for the cells where the selected crops were present, (iii) water limitation for rice was estimated for rainfed crops and water, (vi) thermal limitations were obtained with suitable response curve models, and (v) the final weight of limitations on crop production was obtained by adopting a multiplicative approach.

The results in Figure 3 show the strength of global abiotic limitation and substantially agree with the results given by Buchanan et al. [5]. Moreover, the latitudinal distribution of abiotic limitations is shown in Figure 4.

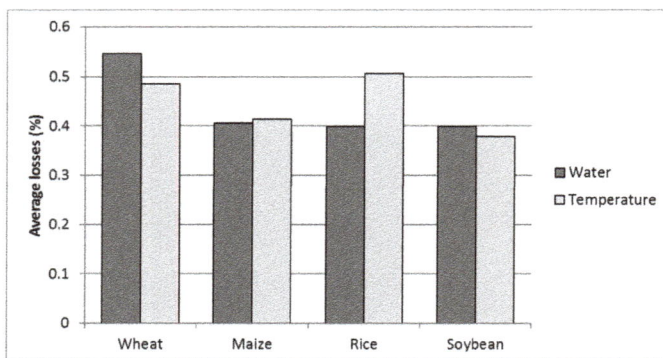

Figure 3. Mean global water and temperature stress losses for the four selected crops, wheat, maize, rice, and soybeans (WMRS) (% on potential net assimilation. PNA). Values above 1 of cumulative water and temperature stresses are the result of non-additive effects of multiple stresses [8].

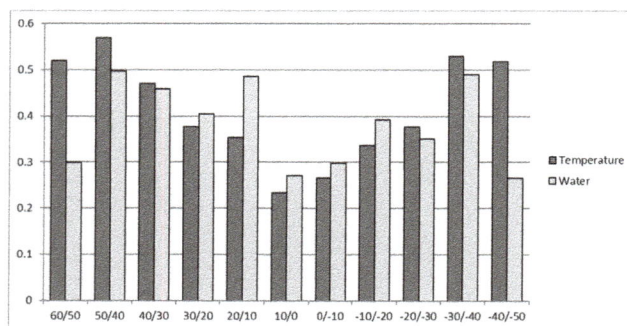

Figure 4. Mean latitudinal water and temperature stress for the four selected crops (WMRS). Values above 1 of cumulative water and temperature stresses are admissible due to the non-additive effects of multiple stresses [8].

Abiotic stresses interact not only among themselves but also with biotic stresses. For example, a crop that has undergone abiotic stress often shows greater susceptibility to attacks from insects, fungi, or mites, and a crop prone to these attacks shows greater sensitivity to water stress because stomata regulation is altered. Oerke [9] provided a global evaluation of the weight of biotic stresses due to weeds, pests, and pathogens for maize and wheat (two staple crops responsible for about 45% of global caloric intake by humans) and cotton (a commodity fundamental for the production of consumer goods). Data from three reference periods (1964–1965, 1988–1990, and 2001–2003) showed weights of 23.9%, 34.0%, and 28.2% for wheat; 34.8%, 38.3%, and 31.2% for maize; and 24.6%, 37.7%, and 28.8% for cotton, respectively.The global weight of biotic stresses on yield losses was estimated to be 70% by Boyer [10] and 13–94% by Farooq et al. [11]. Other data that illustrates both global yield losses and the weight of abiotic stresses have been reported by Cramer [4]. All the above mentioned estimates state the relevant gap between potential and real crop production induced by both biotic and abiotic stresses.

3. Soil Texture, Structure, and Field Hydraulic Arrangements

Plant resilience against abiotic stresses is at a maximum if soil conditions are suitable for plant growth and development. Soil is a disperse three-phase system, and a medium-textured soil at a condition considered optimal for plant growth is schematically composed (by volume) of 50% solid matter (textural particles and organic matter) and 50% pore space. The latter is equally divided between 25% liquid (circulating solution) and 25% gas (soil atmosphere) and an increase in liquid is associated with a decrease in gas and vice versa [12].

A suitable presence of oxygen is crucial for growth and deployment of roots that give an easy access to water and nutrients, and a suitable anchorage [13]. So, a relevant deviation from the abovementioned solid–liquid–gas (slg) volume ratio caused by an excess or shortage of water or by pore volume decrease by compaction strongly affects root growth and deployment, thus affecting crop production [14]. Obviously, the response to unsuitable slg volume ratios is strongly influenced by species and varieties, as testified by the case of ruderal weeds like *Plantago* spp. and *Polygonum aviculare* that are able to colonize compacted soils.

In medium- and fine-textured soils, a volume ratio suitable for most crops is generally attained if colloids (clay and organic matter) aggregate solid elements of texture in structural particles with a diameter of 0.3–3 mm, giving rise to a so-called granular structure, with pores characterized by a 50–50 ratio between macropores (large soil pores generally greater than 0.08 mm in diameter, which, after a saturating rainfall or irrigation, are rapidly drained and re-occupied by air) and micropores (small soil pores with diameters less than 0.08 mm that are mainly found within structural aggregates) [15].

As stated by Valentine et al. [16], who worked on 34 farms located in eastern Scotland that represented a wide range of soil types, textures, crop rotation, and farm management practices, root elongation was directly correlated with the volume of large pores (60–300 μm) and inversely correlated with penetrometer resistance. More specifically, root elongation was enhanced by low-resistance macropores, which overcome mechanical impedance due to the strength of the bulk soil and are limited by hypoxia (or some combination of hypoxia and soil strength) if the rate of oxygen diffusion to the root surface is too low.

Degradation of the granular structure towards structural states less favorable to crops (columnar, blocky, prismatic, massive, etc.) can result from various natural and man-made factors such as the effect of heavy rain on naked soil or the effect of preparatory or tillage carried out under unfavorable conditions of soil water content. Particularly damaging are field activities carried out with excessive soil water content. This explains, at least in part, the importance of hydraulic agricultural systems that aim to avoid excessive water due to rainy or saturated irrigation or high ground.

A fundamental presupposition to reach and maintain the granular structure is to avoid soil water excess; this can be attained by means of suitable soil field hydraulic arrangements. The basic rules for the field management of precipitation are that (1) soil water reservoir must be refilled until field capacity, (2) water excess must be quickly removed from field because it is harmful to most crops, and (3) the speed of this removal must be compatible with the need to avoid harmful erosion phenomena. Field hydraulic arrangements (ditches and drainage systems) should follow these general rules. According to Bonciarelli [17], primary field ditches should be sized with reference to the heavy and frequent rainfall that was quantified for Italy at 50/70 mm in 24 h, which means a reference volume for primary field ditch volumes of 250 to 350 $m^3 \cdot ha^{-1}$. This basic volume should be significantly reduced for very permeable soils. A quantitative analysis carried out by Mariani et al. [18] on rainfall data of 98 stations from throughout Italy, which mainly belongs to the Köppen-Geiger climate types Csa and Cfa, taking into account the 8th absolute 24 h rainfall maximum for the 1993–2012 period, highlighted that most of the territory needed a ditch volume of 250 to 350 $m^3 \cdot ha^{-1}$, as indicated by Bonciarelli [17], with some significant exceptions.

Table 2 lists factors that determine the conservation and improvement of granular structures. Minimum tillage and no tillage aim to promote the self-healing capacity of soils [19–21] and are particularly effective in soils with a sufficient content of good-quality clays.

A significant improvement of structure is also observed with soil organic or inorganic mulching [22] and with soil tillage carried out at a moisture content that maximizes the number of small aggregates [23]. Granular structure can be maintained or enhanced by amendments with organic matter like manure, slurry, compost, crop residues [24,25], or macromolecule polymers acting as soil conditioners [26,27]; or by mixing two or more soil layers in order to reach a more equilibrated texture [28]. Amendments improve soil physical properties, including increasing the content of water-stable aggregates, improving soil porosity and soil penetrability, improving water retention, decreasing soil bulk density and evaporation, and decreasing runoff amount and velocity.

Significant effects on soil structure are played by different tillage systems (moldboard ploughing, minimum tillage, or no tillage systems) with significant effects on macro- and microporosity. For example, Pagliai et al. [19] showed that minimum tillage significantly increased macroporosity, giving rise to a higher hydraulic conductivity and a less pronounced tendency to form a surface crust. Similar effects were highlighted for tropical soils under a no tillage system for 12 years [21].

Soil degradation factors are also listed in Table 2. Soil degradation due to rain [29] first involves the splash erosion that occurs when raindrops hit bare soil and is followed by runoff with sheet erosion (soil removal in thin layers by shallow surface flow), rill erosion (shallow drainage lines less than 30 cm deep), gully erosion (channels deeper than 30 cm that cannot be removed by normal cultivation), and tunnel erosion, which occurs when surface water moves into and through dispersive subsoils [30,31].

Heavy traffic of agricultural machinery is responsible for surface and subsurface soil compaction [28,32], which can be prevented by the adoption of machinery tracked or wheeled with low-pressure tires.

A relevant soil degradation factor is given by freezing–thawing cycles [33], such as the triggering factor for the peculiar type of gully erosion (named "calanchi" in Italian) that is typical of the Apennine clayey hills. The "calanchi" are more frequent in slopes exposed to the south, which are prone to freezing–thawing cycles.

Table 2. List of factors that determine conservation and improvement or degradation of granular soil structure.

Structure Conservation and Improvement Factors	Structure Degradation Factors
Self-healing capacity of soils [20] Organic or inorganic soil mulching [22] Minimum tillage and no tillage [19,21] Tillage at moisture content at which the largest number of small aggregates is produced [23] Amendments: - organic matter (e.g., manure, slurry, crop residues) [24,25] - soil conditioners (polymers) [26,27] - mixing of two or more soil layers in order to reach a more equilibrated texture [28]	Impact of rain and irrigation drops on bare soil surface [29] Runoff effect of rain and irrigation [29] Heavy traffic of agricultural machinery with high-volume pneumatics [28,32] Clods' exposure to freezing–thawing cycles [33]

4. The Impacts of Individual Stress Factors on Crops

4.1. Hypoxia/Anoxia Stress

Crops exposed to limited oxygen conditions must modify their physiology, biochemistry, and transcript profiles to adapt to stressful environments. Crop adaptability can allow survival if exposed to extreme environments or adverse seasons. Several physiological pathways are modified or activated, and many others are repressed [34], to allow plant survival. Specific stress activates a target gene or cluster of genes that may work as signals for cascade activation events and secondary responses [35]. Hypoxia or anoxia is quite a common event that can occur during plant life. During rainy seasons, plants can undergo long periods of flooding and suffer from waterlogging. However, waterlogging can also occur after excessive irrigation in soils with poor drainage. Crops under flooding conditions suffer from low oxygen availability at the root level, which causes a reduction of oxygen in tissues and leads to hypoxia/anoxia responses. In agriculture, oxygen limitation can cause a reduction in crop yield [36]. In fact, yields can dramatically decline after a long period of waterlogging: damages depend on adaptability to hypoxic conditions, soil properties, and drainage. In general, surviving oxygen deprivation depends on which plant tissue type is involved, the developmental stage, and the genotype, as well as the severity and duration of the stress light levels and temperature [37].

The agronomic strategies that can be applied to avoid flooding during the winter provide adequate drainage and soil arrangements, especially where there are structural problems [38]. Another important strategy is nitrate supply. Nitrate fertilization is quite uncommon in winter as plants have low metabolic activity and the frequent rain can increase the nitrate leaching. However, it has been proven that nitrate supply increased plant survival during the winter. Subsequently, several experiments demonstrated that increasing nutrient supply in waterlogged wheat increased plant growth and performance [39]. Among the different nutrients, the most important was nitrogen and, in particular, nitrate. In experiments on tomatoes, plants exposed to oxygen deprivation had delayed anoxia symptoms if nitrate was supplied [40]. The main role is not played by nitrate but the nitrate reductase enzyme. This enzyme in hypoxia conditions has been demonstrated to be involved in NO formation, which plays an important role in the hemoglobin oxygenation/reduction cycle [41].

4.2. Drought Stress and Dry Farming

Drought can drastically reduce crop productivity, especially if it occurs in the most critical stage of plant development. The crop tolerance to water stress depends on the ability of the plant to undergo physiological, biochemical, and morphological changes to enhance water use efficiency (WUE). Crop physiology is regulated by soil water availability and environmental conditions. With optimal soil water availability, plant transpiration is regulated by environmental variables around the leaves. The reduction of water in soil induces in the plant a regulation of the transpiration rate by reduction of stomatal conductance to equilibrate the amount of water uptake and maintain the crop water balance. It means that crops absorb and transpire the same amount of water; in this situation, soil water availability defines the crop yield. The prolonged decline of water availability induces the plant to produce compounds that enhance crop tolerance. Water movement occurs along gradients of water potentials so during drought conditions, crop plants accumulate osmolytes that are used for cell osmoregulation or osmotic adjustment that maintain water uptake [42]. Plants adapted to low water availability also show (i) morphological changes such as reduced leaf area and increased root biomass for exploring a wide volume of soil to find water, and (ii) phenological changes due to the need to reach maturity or ripening as the primary goal. The reduction of water availability induces several physiological and metabolic changes that lead plants to invest their energy to modify their morphology or produce osmolytes, but that reduce yield.

Only a quarter of land receives enough rain to meet crop water requirements. This explains the birth [43] of dry farming, an ensemble of cropping practices that can be adopted in areas without irrigation and where the annual average precipitation is between about 250 to 500 mm, or where rainy events are highly discontinuous and concentrated in limited periods of the year, as for example in Mediterranean environments (Csa type of Köppen-Geiger climate classification), where less than 30% of yearly precipitation falls during the summer. In the Mediterranean, water availability during the summer is often the main limiting factor for agriculture. The water shortage reduces yields and production can be only achieved by using efficient agronomic water management strategies, while climate variability and change have resulted in more sustainable management of water resources.

Dry farming aims in particular at the conservation of water resources by enhancing the storage of useful water in the soil reservoir and limiting water consumption. These objectives can be achieved by increasing the storage of water useful for crops in the soil reservoir, increasing the depth of the soil layer explored by roots, reducing water loss from the soil reservoir, and improving the crop Water Use Efficiency—WUE (water consumption per unit of dry matter produced).

Increased water storage for crops in the soil reservoir can be obtained in the following ways:

- adopting hydraulic arrangements that slow down the speed of surface water runoff and allow water penetration into soil, refilling the underground water reservoir. The infiltration of surface water can also be enhanced by appropriate tillage (e.g., by surface tillage that increases the roughness of the soil). Different tillage methods have a different effect on soil porosity and water infiltration. Ploughing to the depth of 20 cm is the tillage system that gives the highest porosity and water storage capacity, while the lowest was obtained in no-tillage systems [44];
- enhancing the drainage of infiltrated water in order to reach the entire profile explored by roots (e.g., by deep tillage carried out with rippers);
- favoring seasonal flooding of soils by water courses;
- catching runoff water and directing it towards compliant areas suitable for agriculture.

The depth of the layer explored by the roots can be increased in the following ways:

- breaking waterproof or compacted layers by means of ripping or ploughing;
- intervening with drainage or filled ditches to contain the winter rise of the water table, which imposes on crops a superficial root system, enhancing their sensitivity to summer drought.

The soil reservoir can be enhanced by:

- adopting management practices useful for promoting a granular structure with a good equilibrium between macro- and microporosity;
- increasing soil organic matter with organic fertilizers or green manure (the positive effect of organic matter on the soil reservoir is more significant in sandy soils or clay soils with low-quality clays—kaolinite). Organic matter increases the water storage ability of soils. Therefore, higher organic matter concentration means higher water availability for a crop [45];
- adjusting soil texture by mixing surface horizons with excess of sand with lower layers richer in clay (the presupposition for this activity is a suitable analysis of the soil profile and horizon distribution);
- adopting fallow techniques (field plowed and harrowed but left unseeded for one year) aimed at accumulating water in the soil during the "rest" period. For example, the biennial rotation fallow—wheat can be a solution for crop areas where yearly rainfall is insufficient for continuous cultivation. In this context, while the naked fallow should be avoided due to erosion problems and negative impacts on humus content, the most rational form of fallow is that of early autumn plowing (before the rainy season) and superficial work (harrowing) during the following spring and summer, whenever the soil appears encrusted or covered by weeds that waste a relevant quantity of water.

The loss of water from the soil reservoir can be limited by:

- minimizing tillage works (minimum tillage, sod seeding) in order to limit water evaporation from clods exposed to air. Superficial tillage usually limits water losses through evaporation since the capillarity is interrupted and water does not reach the soil surface. Water evaporation can be 70% higher in untilled soils compared with conventional ones [45].

- breaking the soil surface layer or soil crust by means of light soil work (harrows or weeders) in order to eliminate soil cracks that increase the exchange surface with the atmosphere and to interrupt the continuity between soil and atmosphere, reducing water flow towards this latter.
- implementing rational control of weeds, which are strong competitors for water. This is very important during the early stages of the crop cycle and the most critical phases in terms of the water deficit;
- reducing evaporation and transpiration loss through shading, windbreaks, mulching, and anti-hail nets. For example, on tendon-grapevines in Apulia (Italy), plastic films and/or anti-hail nets are used as cover in order to anticipate or delay the harvest and to reduce respiratory losses thanks to the shading effect and the limitation of the transpirational flow. Mulching with plastic or biodegradable films is commonly used in vegetable production to reduce water losses by evaporation and irrigation requirements [46,47]. A reduction of 45% of water need has been demonstrated with combined drip irrigation and mulching, in comparison with overhead sprinkler systems [48].
- using antitranspirants (mostly restricted to nurseries, to avoid excessive transpiration in newly transplanted crops). Antitranspirants are wax or plastic compounds that create a film on the leaves, covering the stomata. The effect is the reduction of water losses by transpiration and the reduction of photosynthesis, which means lower water use and improved tolerance of crops to drought stress [49]. Recent studies report that an antitranspirant sprayed on soybeans under a regular or low irrigation rate was able to improve WUE, acting on stomata and leaf gas exchange [50].
- adopting increased distances between rows and along the row in sowing and transplanting, reducing plant density and competition among plants;
- performing pruning or leaf removal in order to reduce the leaf area index;
- implementing a rational use of fertilizers. In this context, organic fertilization is generally useful for positive effects on soil water retention, and phosphate fertilization is often useful because it stimulates radicle growth, while nitrogen fertilization should be limited to avoid increasing concentrations of the circulating solution with greater difficulty in water supply for plants;
- delivering the water supply strictly needed to restore the useful water soil reservoir;
- choosing more efficient irrigation techniques (considering efficiency to be the ratio between water transpired by a crop to water distributed into the field, the mean efficiency is about 80–90% for drip, subsurface, center pivot or linear irrigation, 60–70% for sprinkler irrigation, and 30–40% for surface irrigation).
- selecting species and varieties so that the stage of maximum sensitivity to water stress does not coincide with the period of maximum dryness for the selected environment;
- choosing early sowing that enhances deep soil colonization by roots and in some species/varieties anticipates harvest. In this sense, autumn sowing is preferable for winter crops (wheat, barley, oat, canola, etc.) while early spring sowing is preferable for summer crops. This choice is obviously suitable only for zones that are not prone to frost risk or where early sowing is compatible with the harvest of previous crops;
- using biostimulants that can improve root development or enhance the biosynthesis of osmotic compounds [51]. These metabolites are able to improve crop tolerance and include plant hormones (abscisic acid) proline, sugars, amino acids, etc. The application of biostimulants can be carried out by soil drench or spray.

WUE can be enhanced by a suitable choice of crop species and varieties. An example is soft wheat (*Triticum aestivum* L.), in which the selection of varieties for the European environment led to varieties suitable for Mediterranean or transitional climates (Csa and Cfa of the Köppen-Geiger climate classification) with the length of the flowering–ripening period reduced with respect to that of varieties suitable for Oceanic environments (Cfb of Köppen-Geiger). Other examples are grapevine rootstocks from *Vitis rupestris* Scheele, which are more resistant than those of *Vitis Riparia* Michx., and sorghum (*Sorghum bicolor* L.), which in environments prone to drought is preferred to maize (*Zea mays* L.) because it is able to resume vegetation without excessive production damage after a drought event.

4.2.1. Precision Farming and Variable-Rate Irrigation

Soil water content is crucial for managing irrigation and can be measured by means of suitable sensors [52,53] or estimated by water balance models based on the continuity equation (conservation of water applied to the soil reservoir). These technologies allow for selecting irrigation time and volumes.

In recent years, these tools have been used to drive variable-rate irrigation (VRI) [54], which allows for the distribution of different amounts of water in different parts of the same field as a function of soil and crop heterogeneity. This approach can be particularly important for fields characterized by a high variability of soil characteristics (texture, structure, depth, and fertility). In these situations, the combination of precision farming and VRI can reduce water losses and improve the WUE of field crops.

4.2.2. Grafting as an Agronomic Tool to Improve Drought Tolerance

WUE can also be improved by means of suitable rootstocks The increase of WUE in plants by grafting can be a reliable agronomic strategy for enhancing crop adaptation and performance in dry environments and, in the past, grafting was widely used in vegetable crops to limit the effects of soil pathogens [55]; more recently, it has been used to induce tolerance against abiotic stresses, such as organic pollutants [56], boron and salinity [57,58], and thermal and water stress. About this latter, it has been shown that the grafting of scions susceptible to water stress onto tolerant rootstocks increased the resistance of grafted trees to this stress [59,60]. Sanders and Markhart [61] have shown that the osmotic potential of dehydrated scions of grafted beans (*Phaseolus vulgaris* L.) was determined by the rootstocks, while the osmotic potential of non-stressed scions was governed by the shoot [60]. Drought tolerance, provided by either the rootstock or the scion, resulted in enhanced nitrogen fixation in soybeans (*Glycine max* L.) [61]. Other grafting experiments on the effect of drought on fruit crops, such as kiwis and grapes, proved that drought-tolerant rootstocks are available and useable under commercial conditions [62,63]. In contrast, only a few studies exist with grafted fruit vegetables. Because eggplants are more effective at water uptake than tomato root systems, it could be useful to study the effects of their grafting on WUE under water-stress conditions. Grafting mini-watermelons onto a commercial rootstock (PS 1313: *Cucurbita maxima* Duchesne × *Cucurbita moschata* Duchesne) revealed a >60% higher marketable yield when grown under conditions of deficit irrigation compared to non-grafted melons [64].

In response to water stress, the hormone that serves as a link between the rootstock and scion seems to be abscisic acid (ABA), as observed in different species such as *Citrus* sp. [65] and *Vitis* sp. [66], but also in cucumber grafts on luffa (*Luffa* sp. Mill. 1754) [67]. Grafting experiments with ABA-deficient mutants of tomato showed that stomata can close independently of the leaf water status, suggesting that there is a chemical signal produced by the roots that controls stomata conductance [68]. Therefore, the selection of rootstocks with adequate biosynthesis and perception of the ABA could improve the efficient use of water and the drought tolerance of many horticultural varieties. Since Cucurbita is one of the genera most used as rootstocks of different species of Cucurbitaceae (watermelon, melon, cucumber), the identification of mutants with a greater ability to synthesize ABA, or hypersensitivity to ABA, could improve tolerance to plant water stress. This increase in ABA biosynthesis also occurred in ethylene-insensitive mutants [69]. On the other hand, it is known that ABA function in the closure of the stomata is mediated by Reactive Oxygen Species (ROS) production [70]. Therefore, used as rootstocks, ethylene-insensitive and ROS-tolerant mutants could also improve water stress tolerance of vegetables, including post-harvest dehydration of fruit.

4.3. Salinity Stress

Soil salinity is determined by the accumulation of soluble salts, which mainly include Cl^-, SO_4^{2-}, HCO_3^-, Na^+, Ca^{2+}, and Mg^{2+}. The accumulation of these ions derives from low-quality irrigation water and poor soil drainage [71]. Salinity reduces plant growth and yield when the concentration of salts reaches 4 dS/m. The reduction of growth and yield depends on crops' tolerance (Table 3).

Table 3. Tolerance thresholds expressed as electric conductivity (EC) and critical EC values for yield loss. Tolerance degree expressed as: S = sensitive; MS = moderately sensitive; MT = moderately tolerant; T = tolerant [72].

Vegetable Crops	Soil Salinity			Salinity in Irrigation Water			Tolerance
	Threshold (CEe) (dS·m^{-1})	Slope (%/dS·m^{-1})	Yield 0% (dS·m^{-1})	Threshold (CEi) (dS·m^{-1})	Slope (%/dS·m^{-1})	Yield 0% (dS·m^{-1})	
Artichoke (*Cynara scolymus* L.)	4.8	10.9	14	2.7	14.4	9.6	MT
Asparagus (*Asparagus officinalis* L.)	4.1	2	54.1	2.7	3.0	36	T
Bean (*Faseolus vulgaris* L.)	1	19	6.3	0.7	28.5	4.2	S
Broad bean (*Vicia faba* L.)	1.6	9.6	12	1.1	14.5	8	MS
Broccoli (*Brassica oleracea* var. *italica* Plenck)	2.8	9.2	13.7	1.9	13.8	9.2	MS
Brussels sprouts (*Brassica oleracea* var. *gemmifera* DC.)	-	-	-	-	-	-	MS
Cabbage hood (*Brassica oleracea* var. *capitata* L.)	1.8	9.7	12.1	1.2	14.6	8.1	MS
Carrot (*Daucus carota* L.)	1	14	8.1	0.7	21.0	5.5	S
Cauliflower (*Brassica oleracea* L. var. *botrytis* L.)	-	-	-	-	-	-	MS
Celery (*Apium graveolens* L. var. *dulce* [Mill.] Pers.)	1.8	6.2	17.9	1.2	9.3	12	MS
Cowpea (*Vigna unguiculata* [L.] Walpers subsp. *unguiculata*)	4.9	12	13.2	3.3	18.2	8.8	MT
Cucumber (*Cucumis sativus* L.)	2.5	13	10.2	1.7	19.5	6.8	MS
Eggplants (*Solanum melongena* L.)	1.1	6.9	15.6	0.7	10.3	10.4	MS
Funnel (*Foeniculum vulgare* Miller var. *azoricum* [Mill.] Thell.)	1.5	15	8.2	1.1	18.0	6.7	MS
Garlic (*Allium sativum* L.)	1.7	10	11.7	1.1	14.9	7.8	MS
Lettuce (*Lactuca sativa* L.)	1.3	13	9	0.9	19.5	6	MS
Melon (*Cucumis melo* L.)	1	8.4	12.9	0.7	12.7	8.6	MS
Onion (*Allium cepa* L.)	1.2	16	7.5	0.8	24.0	5	S
Pea (*Pisum sativum* L.)	-	-	-	-	-	-	S
Pepper (*Capsicum annuum* L.)	1.5	14	8.6	1.0	21.0	5.8	MS
Potato (*Solanum tuberosum* L.)	1.7	12	10	1.1	18.0	6.7	MS
Radish (*Raphanus sativus* L.)	1.2	13	8.9	0.8	19.5	5.9	MS
Rapa (*Brassica rapa* L. var. *rapa*)	0.9	9	12	0.7	13.5	8.1	MS
Spinach (*Spinacia oleracea* L.)	2	7.6	15.2	1.3	11.4	10.1	MS
Straberry (*Fragaria x ananassa* Duch.)	1	33	4.0	0.7	49.5	2.7	S
Swiss chard (*Beta vulgaris* L. var. *conditiva* Alef.)	4	9	15.1	2.7	13.5	10.1	MT
Tomato (*Lycopersicon esculentum* Mill.)	2.5	9.9	12.6	1.7	15.0	8.4	MS
Water melon (*Citrullus lanatus* [Thunberg] Matsumura et Nakai)	-	-	-	-	-	-	MS
Zucchini (*Cucurbita pepo* L.)	4.7	9.4	15.3	3.1	14.1	10.2	MT

High salinity is often a problem in areas located along the sea, especially in Mediterranean areas with intensive agriculture [73] and, in particular, for vegetable farms. Vegetable crops usually have short production cycles and require substantial amounts of water in short periods, increasing salinity problems. The majority of high salinity soil and irrigation water occurs in summer because the reduction of water availability increases seawater infiltration in the groundwater. In these conditions farms that use underground water pumped from soils for irrigation increase seawater infiltration. It has been estimated that half of irrigated agricultural lands are affected by salinity.

Agronomic strategies to reduce salinity stress during cultivation can act on soil or crops. At the soil level, the simplest strategy is to maintain high water availability, with frequent irrigation if possible. It is important to use irrigation systems with high efficiency, such as drip irrigation. The reduction of soil water content increases the salt concentration and crops suffer from osmotic stress. Summer is the most critical period for salinity stress; usually increased plant survival can be obtained by calcium nitrate or chloride [74]. The application of calcium has a beneficial effect on both the soil structure and plant tolerance. The calcium ions move the sodium ions from soil colloids and these can be leached by irrigation. The same effect can be reached by magnesium application. The mitigation effect is considered in the calculation of the sodium adsorption rate (SAR), which takes into consideration the concentration of sodium, calcium, and magnesium, in the following equation:

$$\text{SAR} = \frac{Na^+}{\sqrt{\frac{Ca^{2+} + Mg^{2+}}{2}}}. \tag{1}$$

Therefore, fertilizers containing calcium and magnesium in sodic soils improve the structure of soils and provide a better environment for roots and plant growth. At the crop physiology level, cytosolic calcium inhibits sodium channels in membranes, called salt overlay sensitivity (SOS), and reduces salt accumulation in the cells, alleviating salinity stress. The mutation of a gene encoding for a SOS plasma membrane Na^+–H^+ antiporter increased salt sensitivity, while the overexpression of this gene increased salt tolerance [75]. Moreover, nitrates, if calcium nitrate is used, are in competition with sodium ions for accumulation in the vacuoles. Therefore, the supply of nitrates may reduce salt uptake. However, this particular aspect needs further investigation.

Positive effects have been reported for the application of plant growth-promoting bacteria (PGPB) in increasing crop tolerance to salinity [76]. The mechanism of action has not been elucidated yet but it seems that the bacteria help the roots avoid the excessive uptake of sodium. An analogous effect can be obtained using arbuscular-mycorrhizal fungi, which can improve the uptake of mineral nutrients and reduce salt stress, enhance osmotic adjustment, and have a direct effect on plant hormone biosynthesis. The application of *Glomus* species in lettuce stressed with sodium chloride improved photosynthetic activity, stomatal conductance, and WUE [77].

4.4. Lodging

The process by which shoots of winter or summer cereals are displaced from their vertical orientation is named lodging. In cereals such as wheat and barley, lodging is most likely to occur during the two or three months preceding harvest, usually after ear or panicle emergence, with the result that shoots permanently lean or lie horizontally on the ground. Lodging can be caused by the buckling of stems (stem lodging) or displacement of roots within the soil (root lodging) [78]. In stem lodging, roots are held firm in a strong soil where the wind force buckles one of the lower internodes of the shoot. Root lodging becomes more likely when the anchorage strength is reduced by weak soil or poorly developed anchorage roots. The effect is a reduction in crop yield by up to 80%, with further losses in grain quality, greater drying costs, and an increase in the time taken for harvesting [79].

The main factors that contribute to the lodging process are strong winds (e.g., foehn winds or downbursts associated with thunderstorms), heavy rain, crop pests (e.g., *Diabrotica vergifera* larvae bore deep into the roots, destroying them and giving rise to root lodging), diseases (e.g., fungal diseases that attack the basal part the stems of winter cereals give rise to stem lodging), and an excess of nitrogen fertilization (nitrogen enhances the vegetative growth of stems, with excessive elongation associated with a lower elasticity and increased weakness).

Lodging can be prevented by installing shelterbelts against dominant winds, adopting an equilibrated nitrogen fertilization, choosing a suitable plant density, and optimizing the management of pests and biotic diseases harmful to the stability and mechanic resistance of stems and roots. The lodging risk for crops can also be reduced by the introduction of semi-dwarf varieties (e.g., the wheat varieties produced by Nazareno Strampelli and Norman Borlaug in the 20th century) or by the adoption of plant growth regulators (PGRs).

Wind breaks are barriers given by plantations (trees and shrubs) or non-living material (walls, fences and so all), established in order to protect field crops from dominant winds. A crucial decision in order to optimize the effect of windbreaks is influenced by their porosity, because a very dense or low-density row of trees have low effectiveness, while the most effective are medium-density rows [80]. Numerous studies describe the effect of wind breaks on various atmospheric variables such as temperature (for example, the risk of frost can be increased) and evapotranspiration. Furthermore, shelter belt plantations show a more or less strong competition with crops for light, water, and nutrients [64]. All these effects, which are generally a function of the distance from the windbreak, should be considered in the design of these artifacts.

5. Conclusions

This review discussed agronomic strategies that can be adopted to cope with the effects of abiotic stress on crops, offering a series of ideas based on suitable cultivation techniques. Often defense against abiotic stress is only sought at the genetic level by the identification of tolerant genotypes. This is a correct approach, but agronomic tools can often offer an adequate and rapid solution for reducing crop yield losses. The interaction between genetics and management was a crucial factor of the 20th-century green revolution and is destined to receive increasing attention in the coming years due to the need to increase global agricultural production while respecting the quality requirements of the market. Agronomic management strategies have been considered static for a long time and have not been adequately reconsidered for controlling crop performance. Instead, agronomic management has to be continuously revised, considering innovations in crop tolerance and genetic improvements.

Author Contributions: Both authors contributed equally to the review.

Conflicts of Interest: The authors declare no conflict of interest.

References

1. Zandalinas, S.I.; Mittler, R.; Balfagón, D.; Arbona, V.; Gómez-Cadenas, A. Plant adaptations to the combination of drought and high temperatures. *Physiol. Plant.* **2017**. [CrossRef] [PubMed]
2. Vinocur, B.; Altman, A. Recent advances in engineering plant tolerance to abiotic stress: Achievements and limitations. *Curr. Opin. Biotechnol.* **2005**, *16*, 123–132. [CrossRef] [PubMed]
3. Cramer, G.R.; Urano, K.; Delrot, S.; Pezzotti, M.; Shinozaki, K. Effects of abiotic stress on plants: A systems biology perspective. *BMC Plant Biol.* **2011**, *11*, 163. [CrossRef] [PubMed]
4. Atkinson, N.J.; Urwin, P.E. The interaction of plant biotic and abiotic stresses: From genes to the field. *J. Exp. Bot.* **2012**, *63*, 3523–3543. [CrossRef] [PubMed]
5. Buchanan, B.B.; Gruissem, W.; Russell, L.J. (Eds.) *Biochemistry and Molecular Biology of Plants*, 2nd ed.; Wiley: Hoboken, NJ, USA, 2015; 1280p, ISBN 978-0-470-71421-8.
6. Boogaard, H.; Wolf, J.; Supit, I.; Niemeyer, S.; van Ittersum, M. A regional implementation of WOFOST for calculating yield gaps of autumn-sown wheat across the European Union. *Field Crops Res.* **2013**, *143*, 130–142. [CrossRef]
7. Singh, R.; Van Dam, J.C.; Feddes, R.A. Water productivity analysis of irrigated crops in Sirsa district, India. *Agric. Water Manag.* **2006**, *82*, 253–278. [CrossRef]
8. Mariani, L. Carbon plants nutrition and global food security. *Eur. Phys. J. Plus* **2017**, *132*, 69. [CrossRef]
9. Oerke, E.C. Crop losses to pests. Centenary review. *J. Agric. Sci.* **2006**, *144*, 31–43. [CrossRef]

10. Boyer, J.S. Plant productivity and environment. *Science* **1982**, *218*, 443–448. [CrossRef] [PubMed]

11. Farooq, M.; Wahid, A.; Kobayashi, N.; Fujita, D.; Basra, S.M.A. Plant drought stress: Effects, mechanisms and management. *Agron. Sustain. Dev.* **2009**, *29*, 185–212. [CrossRef]

12. Hillel, D. *Introduction to Environmental Soil Physics*; Elsevier: Amsterdam, The Netherlands, 2013; 495p.

13. Rich, S.M.; Watt, M. Soil conditions and cereal root system architecture: Review and considerations for linking Darwin and Weaver (Darwin review). *J. Exp. Bot.* **2013**, *64*, 1193–1208. [CrossRef] [PubMed]

14. Passioura, J.B. Soil conditions and plant growth. *Plant Cell Environ.* **2002**, *25*, 311–318. [CrossRef] [PubMed]

15. Pearson, C.J.; Norman, D.W.; Dixon, J. *Sustainable Dryland Cropping in Relation to Soil Productivity—FAO Soils Bulletin 72*; Food & Agriculture Organization of the United Nations (FAO): Rome, Italy, 1995.

16. Valentine, T.A.; Hallett, P.D.; Binnie, K.; Young, M.W.; Squire, G.R.; Hawes, C.; Bengough, A.G. Soil strength and macropore volume limit root elongation rates in many UK agricultural soils. *Ann. Bot.* **2012**, *110*, 259–270. [CrossRef] [PubMed]

17. Bonciarelli, F. *Fondamenti di Agronomia Generale*; Edagricole: Bologna, Italy, 1989; 292p.

18. Mariani, L.; Cola, G.; Parisi, S. Dimensioning of field ditches in function of heavy and frequent precipitations. In Proceedings of the AIAM 2013 17th Annual Meeting of the Italian Agrometeorological Association Agrometeorology for Environmental and Food Security, Florence, Italy, 4–6 June 2013. (supplement to the *Italian Journal of Agrometeorology* 2013, 101–102).

19. Pagliai, M.; Vignozzi, N.; Pellegrini, S. Soil structure and the effect of management practices. *Soil Tillage Res.* **2004**, *79*, 131–143. [CrossRef]

20. Kibblewhite, M.G.; Ritz, K.; Swift, M.J. Soil health in agricultural systems. *Philos. Trans. R. Soc. Lond. B Biol. Sci.* **2008**, *363*, 685–701. [CrossRef] [PubMed]

21. Carmeis Filho, A.C.A.; Crusciol, C.A.C.; Guimarães, T.M.; Calonego, J.C.; Mooney, S.J. Correction: Impact of amendments on the physical properties of soil under tropical long-term no till conditions. *PLoS ONE* **2017**, *12*. [CrossRef] [PubMed]

22. Mulumba, N.L.; Lal, R. Mulching effects on selected soil physical properties. *Soil Tillage Res.* **2008**, *98*, 106–111. [CrossRef]

23. Wagner, L.E.; Ambe, N.M.; Barnes, P. Tillage-induced soil aggregate status as influenced by water content. *Trans. ASAE* **1992**, *35*, 499–504. [CrossRef]

24. Liu, Z.; Rong, Q.; Zhou, W.; Liang, G. Effects of inorganic and organic amendment on soil chemical properties, enzyme activities, microbial community and soil quality in yellow clayey soil. *PLoS ONE* **2017**, *12*, e0172767. [CrossRef] [PubMed]

25. Tejada, M.; Gonzalez, J.L. Influence of organic amendments on soil structure and soil loss under simulated rain. *Soil Tillage Res.* **2017**, *93*, 197–205. [CrossRef]

26. Wu, S.F.; Wu, P.T.; Feng, H.; Bu, C.F. Influence of amendments on soil structure and soil loss under simulated rainfall China's loess plateau. *Afr. J. Biotechnol.* **2010**, *9*, 6116–6121.

27. Su, L.; Wang, Q.; Wang, C.; Shan, Y. Simulation Models of Leaf Area Index and Yield for Cotton Grown with Different Soil Conditioners. *PLoS ONE* **2015**. [CrossRef] [PubMed]

28. Pearson, C.J.; Norman, D.W.; Dixon, J. Physical aspects of crop productivity. In *Sustainable Dryland Cropping in Relation to Soil Productivity, FAO Soils Bulletin 72*; Food & Agriculture Organization of the United Nations (FAO): Rome, Italy, 1995; Chapter 2; Available online: http://www.fao.org/docrep/V9926E/v9926e04.htm (accessed on 22 September 2017).

29. Mohamadi, M.A.; Kavian, A. Effects of rainfall patterns on runoff and soil erosion in field plots. *Int. Soil Water Conserv. Res.* **2015**, *3*, 273–281. [CrossRef]

30. NSW DPI Soils Advisory Office. Soil Erosion Solutions, Fact Sheet 1: Types of Erosion. 2017. Available online: https://www.dpi.nsw.gov.au/__data/assets/pdf_file/0003/255153/fact-sheet-1-types-of-erosion.pd (accessed on 3 September 2017).

31. Taguas, E.V.; Guzman, E.; Guzman, G.; Vanwalleghem, T.; Gomez, J.A. Characteristics and Importance of Rill and Gully Erosion. *Cuadenos de Investigacion Geografica* **2015**, *41*, 107–126. [CrossRef]

32. Soracco, C.G.; Lozano, L.A.; Villarreal, R.; Palancar, T.C.; Collazo, D.J.; Sarli, G.O.; Filgueira, R.R. Effects of compaction due to machinery traffic on soil pore configuration. *Rev. Bras. Ciênc. Solo* **2015**, *39*. [CrossRef]

33. Dagesse, D.F. Freezing cycle effects on water stability of soil aggregates. *Can. J. Soil Sci.* **2013**, *93*, 473–483. [CrossRef]

34. Tennis, E.S.; Dolferus, R.; Ellis, M.; Rahman, M.; Wu, Y.; Hoeren, F.U.; Grover, A.; Ismond, K.P.; Good, A.G.; Peacock, W.J. Molecular strategies for improving waterlogging tolerance in plants. *J. Exp. Bot.* **2000**, *51*, 89–97.

35. Drew, M.C. Oxygen deficiency and root metabolism: Injury and acclimation under hypoxia and anoxia. *Ann. Rev. Plant Physiol. Plant Mol. Biol.* **1997**, *48*, 223–250. [CrossRef] [PubMed]

36. Linkemer, G.; Board, J.E.; Musgrave, M.E. Waterlogging effects on growth and yield components in late-planted soybean. *Crop Sci.* **1998**, *38*, 1576–1584. [CrossRef] [PubMed]

37. Fukao, T.; Bailey-Serres, J. Plant responses to hypoxia—Is survival a balancing act? *Trends Plant Sci.* **2004**, *9*, 449–456. [CrossRef] [PubMed]

38. MacEwan, R.J.; Gardner, W.K.; Ellington, A.; Hopkins, D.G.; Bakker, A.C. Tile and mole drainage for control of waterlogging in duplex soils of south-eastern Australia. *Aust. J. Exp. Agric.* **1992**, *32*, 865–878. [CrossRef]

39. Huang, B.; Johnson, J.W.; Nesmith, S.; Bridges, D.C. Growth physiological and anatomical responses of two wheat genotypes to waterlogging and nutrient supply. *J. Exp. Bot.* **1994**, *45*, 193–202. [CrossRef]

40. Allègre, A.; Silvestre, J.; Morard, P.; Kallerhoff, J.; Pinelli, E. Nitrate reductase regulation in tomato roots by exogenous nitrate: A possible role in tolerance to long-term root anoxia. *J. Exp. Bot.* **2004**, *55*, 2625–2634. [CrossRef] [PubMed]

41. Igamberdiev, A.U.; Hill, R.D. Nitrate, NO and haemoglobin in plant adaptation to hypoxia: An alternative to the classic fermentation pathways. *J. Exp. Bot.* **2004**, *55*, 2473–2482. [CrossRef] [PubMed]

42. Serraj, R.; Sinclair, T.R. Osmolyte accumulation: Can it really help increase crop yield under drought conditions? *Plant Cell Environ.* **2002**, *25*, 333–341. [CrossRef] [PubMed]

43. Widtsoe, J.A. *Dry Farming, a System of Agriculture for Countries under a Low Rainfall*; The Mcmillan Company: New York, NY, USA, 1920; p. 501. Available online: http://archive.org/details/dryfarmingasyst01widtgoog (accessed on 15 September 2017).

44. Lipiec, J.; Kuś, J.; Słowińska-Jurkiewicz, A.; Nosalewicz, A. Soil porosity and water infiltration as influenced by tillage methods. *Soil Tillage Res.* **2006**, *89*, 210–220. [CrossRef]

45. Bond, J.J.; Willis, W.O. Soil water evaporation: Surface residue rate and placement effects. *Soil Sci. Soc. Am. J.* **1969**, *33*, 445–448. [CrossRef]

46. Gupta, S.; Larson, W.E. Estimating soil water retention characteristics from particle size distribution, organic matter percent, and bulk density. *Water Resour. Res.* **1979**, *15*, 1633–1635. [CrossRef]

47. Lament, W.J. Plastic mulches for the production of vegetable crops. *HortTechnology* **1993**, *3*, 35–39.

48. Kasirajan, S.; Ngouajio, M. Polyethylene and biodegradable mulches for agricultural applications: A review. *Agron. Sustain. Dev.* **2012**, *32*, 501–529. [CrossRef]

49. Clough, G.H.; Locascio, S.J.; Olson, S.M. Continuous use of polyethylene mulched beds with overhead or drip irrigation for successive vegetable production. In Proceedings of the 20th National Agriculture Plastics Congress, Portland, OR, USA, 25–27 August 1987; pp. 57–61.

50. Davenport, D.; Hagan, R.; Martin, P. Antitranspirants uses and effects on plant life. *Calif. Agric.* **1969**, *23*, 14–16.

51. Shasha, J.I.; Ling, T.O.N.G.; Fusheng, L.I.; Hongna, L.U.; Sien, L.I.; Taisheng, D.U.; Youjie, W.U. Effect of a new antitranspirant on the physiology and water use efficiency of soybean under different irrigation rates in an arid region. *Front. Agric. Sci. Eng.* **2017**, *4*, 155–164.

52. Bulgari, R.; Cocetta, G.; Trivellini, A.; Vernieri, P.; Ferrante, A. Application of biostimulants for improving yield and quality of vegetables and floricultural crops. *Biol. Agric. Hortic.* **2015**, *31*, 1–17. [CrossRef]

53. Alvino, A.; Marino, S. Remote sensing for irrigation of horticultural crops. *Horticulturae* **2017**, *3*, 40. [CrossRef]

54. Dukes, M.D.; Perry, C. Uniformity testing of variable-rate center pivot irrigation control systems. *Precis. Agric.* **2006**, *7*, 205. [CrossRef]

55. Schwarz, D.; Rouphael, Y.; Colla, G.; Venema, J.H. Grafting as a tool to improve tolerance of vegetables to abiotic stresses: Thermal stress, water stress and organic pollutants. *Sci. Hortic.* **2010**, *127*, 162–171. [CrossRef]

56. Edelstein, M.; Ben-Hur, M.; Plaut, Z. Grafted melons irrigated with fresh or effluent water tolerate excess boron. *J. Am. Soc. Hortic. Sci.* **2007**, *132*, 484–491.

57. Edelstein, M.; Plaut, Z.; Ben-Hur, M. Sodium and chloride exclusion and retention by non-grafted and grafted melon and Cucurbita plants. *J. Exp. Bot.* **2011**, *62*, 177–184. [CrossRef] [PubMed]

58. Garcia-Sanchez, F.; Syvertsen, J.P.; Gimeno, V.; Botia, P.; Perez-Perez, J.G. Responses to flooding and drought stress by two citrus rootstock seedlings with different water-use efficiency. *Biol. Plant* **2007**, *130*, 532–542. [CrossRef]

59. Satisha, J.; Prakash, G.S.; Bhatt, R.M.; Sampath Kumar, P. Physiological mechanisms of water use efficiency in grape rootstocks under drought conditions. *Int. J. Agric. Res.* **2007**, *2*, 159–164.

60. Sanders, P.L.; Markhart, A.H., III. Interspecific grafts demonstrate root system control of leaf water status in water stressed Phaseolus. *J. Exp. Bot.* **1992**, *43*, 1563–1567. [CrossRef]

61. Serraj, R.; Sinclair, T.R. Processes contributing to N_2-fixation insensitivity to drought in the soybean cultivar Jackson. *Crop Sci.* **1996**, *36*, 961–968. [CrossRef]

62. Clearwater, M.J.; Lowe, R.G.; Hofstee, B.J.; Barclay, C.; Mandemaker, A.J.; Blattmann, P. Hydraulic conductance and rootstock effects in grafted vines of kiwifruit. *J. Exp. Bot.* **2004**, *55*, 1371–1381. [CrossRef] [PubMed]

63. Rouphael, Y.; Cardarelli, M.; Colla, G.; Rea, E. Yield, mineral composition, water relations, and water use efficiency of grafted mini-watermelon plants under deficit irrigation. *HortScience* **2008**, *43*, 730–736.

64. Allario, T.; Brumos, J.; Colmenero-Flores, J.M.; Iglesias, D.J.; Pina, J.A.; Navarro, L.; Talon, M.; Ollitrault, P.; Morillon, R.; Morillon, R. Tetraploid Rangpur lime rootstock increases drought tolerance via enhanced constitutive root abscisic acid production. *Plant Cell Environ.* **2013**, *36*, 856–868. [CrossRef] [PubMed]

65. Serra, I.; Strever, A.; Myburgh, P.A.; Deloire, A. The interaction between rootstocks and cultivars (*Vitis vinifera* L.) to enhance drought tolerance in grapevine. *Aust. J. Grape Wine Res.* **2014**, *20*, 1–14. [CrossRef]

66. Liu, S.; Li, H.; Lv, X.; Ahammed, G.J.; Xia, X.; Zhou, J.; Zhou, Y. Grafting cucumber onto luffa improves drought tolerance by increasing ABA biosynthesis and sensitivity. *Sci. Rep.* **2016**, *6*, 20212. [CrossRef] [PubMed]

67. Holbrook, N.M.; Shashidhar, V.R.; James, R.A.; Munns, R. Stomatal control in tomato with ABA-deficient roots: Response of grafted plants to soil drying. *J. Exp. Bot.* **2002**, *53*, 1503–1514. [PubMed]

68. Corbineau, F.; Xia, Q.; Bailly, C.; El-Maarouf-Bouteau, H. Ethylene, a key factor in the regulation of seed dormancy. *Front. Plant Sci.* **2014**, *5*, 539. [CrossRef] [PubMed]

69. Mustilli, A.C.; Merlot, S.; Vavasseur, A.; Fenzi, F.; Giraudat, J. Arabidopsis OST1 protein kinase mediates the regulation of stomatal aperture by abscisic acid and acts upstream of reactive oxygen species production. *Plant Cell* **2002**, *14*, 3089–3099. [CrossRef] [PubMed]

70. Bernstein, L. Effects of salinity and sodicity on plant growth. *Ann. Rev. Phytopathol.* **1975**, *13*, 295–312. [CrossRef]

71. Acosta-Motos, J.R.; Ortuño, M.F.; Bernal-Vicente, A.; Diaz-Vivancos, P.; Sanchez-Blanco, M.J.; Hernandez, J.A. Plant Responses to Salt Stress: Adaptive Mechanisms. *Agronomy* **2017**, *7*, 18. [CrossRef]

72. FAO. Annex 1. Crop Salt Tolerance Data. Available online: http://www.fao.org/docrep/005/y4263e/y4263e0e.htm (accessed on 12 September 2017).

73. Jaleel, C.A.; Manivannan, P.; Sankar, B.; Kishorekumar, A.; Gopi, R.; Somasundaram, R.; Panneerselvam, R. Water deficit stress mitigation by calcium chloride in *Catharanthus roseus*: Effects on oxidative stress. proline metabolism and indole alkaloid accumulation. *Colloids Surf. B Biointerfaces* **2007**, *60*, 110–116. [CrossRef] [PubMed]

74. Yang, Q.; Chen, Z.Z.; Zhou, X.F.; Yin, H.B.; Li, X.; Xin, X.F.; Gong, Z. Overexpression of SOS (Salt Overly Sensitive) genes increases salt tolerance in transgenic Arabidopsis. *Mol. Plant* **2009**, *2*, 22–31. [CrossRef] [PubMed]

75. Mayak, S.; Tirosh, T.; Glick, B.R. Plant growth-promoting bacteria confer resistance in tomato plants to salt stress. *Plant Physiol. Biochem.* **2004**, *42*, 565–572. [CrossRef] [PubMed]

76. Ruiz-Lozano, J.M.; Azcon, R.; Gomez, M. Alleviation of salt stress by arbuscular-mycorrhizal Glomus species in *Lactuca sativa* plants. *Physiol. Plant.* **1996**, *98*, 767–772. [CrossRef]

77. Berry, P.M. Understanding and reducing lodging in cereals. *Adv. Agron.* **2004**, *84*, 217–271.

78. Tams, A.R.; Mooney, S.J.; Berry, P.M. The Effect of Lodging in Cereals on Morphological Properties of the Root-Soil Complex. In Proceedings of the SuperSoil 2004: 3rd Australian New Zealand Soils Conference, Sydney, Australia, 5–9 December 2004. Available online: http://www.regional.org.au/au/asssi/supersoil2004/s9/oral/1998_tamsa.htm (accessed on 20 September 2017).

79. Bean, A.; Alperi, R.W.; Federer, C.A. A method for categorizing shelterbelts porosity. *Agric. Meteorol.* **1975**, *14*, 417–429. [CrossRef]
80. Campi, P.; Palumbo, A.D.; Mastrorilli, M. Effect of tree windbreaks on microcliamte and wheat productivity in a Mediterrranean environment. *Eur. J. Agron.* **2009**, *30*, 220–227. [CrossRef]

MDPI

St. Alban-Anlage 66

4052 Basel

Switzerland

Tel. +41 61 683 77 34

Fax +41 61 302 89 18

www.mdpi.com

Horticulturae Editorial Office

E-mail: horticulturae@mdpi.com

www.mdpi.com/journal/horticulturae

www.ingramcontent.com/pod-product-compliance
Lightning Source LLC
Chambersburg PA
CBHW051912210326
41597CB00033B/6122